FEEDING THE BIRDS AT YOUR TABLE

DARRYL JONES is a Professor of Ecology at Griffith University in Brisbane, where for over 30 years he has been investigating the many ways that people and wildlife interact. He is particularly interested in why some species are extremely successful in urban landscapes while many others are not, and how best to deal with the ensuing conflicts. More recently he has been trying to understand more about the humans that also live in cities in large numbers, and how and why they engage with nature. This has led him into the strange and fascinating world of wild bird feeding and has resulted in collaborations with other researchers all over the world. He has published more than 170 scientific papers and six books, including *The Birds at My Table* (2018).

DEDICATION

For Danielle Hodgson and Cara Parsons.
The best learning and teaching happens in both
directions.

FEEDING THE BIRDS AT YOUR TABLE

A GUIDE FOR AUSTRALIA

DARRYL JONES

NEWSOUTH

A NewSouth book

Published by
NewSouth Publishing
University of New South Wales Press Ltd
University of New South Wales
Sydney NSW 2052
AUSTRALIA
newsouthpublishing.com

© Darryl Jones 2019
First published 2019

10 9 8 7 6 5 4 3 2 1

A catalogue record for this book is available from the National Library of Australia

ISBN   9781742236322 (paperback)
        9781742244594 (ebook)
        9781742249087 (ePDF)

Design Josephine Pajor-Markus
Cover design Madeleine Kane
Cover image Major Mitchell's cockatoos, Leila Jeffreys, www.leilajeffreys.com
Illustrations Michelle Baker, www.michellebaker.org
Printer Griffin Press

All reasonable efforts were taken to obtain permission to use copyright material reproduced in this book, but in some cases copyright could not be traced. The author welcomes information in this regard.

This book is printed on paper using fibre supplied from plantation or sustainably managed forests.

CONTENTS

1

FIRST

WHY WE NEED TO TALK
ABOUT BIRD FEEDING

Why would we possibly need a guide to feeding wild birds in Australia? Surely everyone knows that we shouldn't be doing it in the first place. We have all seen the DO NOT FEED THE WILDLIFE signs at the local pond or on the picnic tables in national parks. For decades this has been the message end-lessly repeated by the RSPCA, wildlife carers, vets, zoos, city councils, state agencies, bird societies and conservation and environment groups – feeding wild animals is bad, irresponsible and sometimes even dangerous. These anti-feeding messages are typically accompanied by alarming claims. Wildlife will become dependent on human-provided foods.

Animals will forget how to forage naturally and will suffer as a result of their inadequate diet. Feeding wildlife assists the spread of disease and attracts vermin and predators. For anyone who cares about our wildlife, these are confronting issues.

Serious, obviously, but it's important to appreciate that feeding wildlife takes two very different forms, and each of these poses different challenges. The signposts and most strident warnings relate mainly to the feeding of wildlife in public spaces. We have probably all witnessed or even directly experienced the behaviour of wild animals that have been given food by humans, and then expect more of it: overly familiar possums or kangaroos at camping grounds; pushy ducks and swans at urban ponds; kamikaze kookaburras joining the barbecue; or the increasingly in-your-face ibis of city parklands. There are some genuinely troubling problems associated with feeding wildlife, and there is clear scientific evidence that some of the birds and other animals in these settings are suffering as a result of being fed too much bad food. Think of the waterbirds consuming all that bread at the duck pond, and the gulls that seem to subsist primarily on chips. We certainly do need to think seriously about whether we should be feeding wildlife in public places, and this is something I will be discussing in chapters 2 and 4.

(Spoiler alert: feeding bread to ducks, and chips to gulls, is never a good idea!)

FOR THE MILLIONS OF AUSTRALIANS WHO REGULARLY FEED WILD BIRDS AT THEIR HOMES, THE PRACTICE CAN BE CALMING, HEALING AND INSTRUCTIVE, OFFERING A CHANCE TO CONNECT WITH NATURE.

However, the primary focus of this book is the type of wildlife feeding that we do at home – in our backyard, or on our verandah or balcony. I will be concentrating almost entirely on birds – although there is plenty of possum, bluetongue, wallaby and even echidna feeding going on as well (we know even less about the effects of feeding on these other types of animals than we do about birds). For the millions of Australians who feed wild birds at their homes (and this includes me), this is almost always something they do regularly, often at the same time every day. It is planned, systematic and organised. The food we provide to our feathered visitors has typically been purchased specifically for that purpose, and the aim is to attract these wonderful – and completely wild – birds to our homes. If the birds take up the offer, these visits can bring real pleasure and meaning to our lives. For many people, this simple activity is profoundly important; it can be calming, healing and instructive, representing a simple opportunity to connect with nature – something that is becoming increasingly hard to do as cities grow and biodiversity retreats.

Obviously, not everyone feeding birds would describe the experience in these terms. But numerous studies from Australia and elsewhere have found that most feeders (the people providing the food, not the feeding platform) look forward to observing their visitors on a regular basis. They also care about these birds and hope that what they are providing is good for them. For many feeders, this is a legitimate worry. Sure, birds eat just about everything that is put out for them. But that does not mean the food is necessarily safe and nutritious. After all, we humans also eat plenty of food that is terrible for us, and we know what happens as a result.

What many people who feed birds have been trying to find – and what I am seeking to provide here – is reliable information and advice on what they should be feeding their birds, and what they should be avoiding. Most dedicated feeders are quite aware of the well-known concerns referred to above, but simply don't know where to find the advice they need.

A search of the internet will reinforce the conclusion that feeding birds in Australia is wrong and possibly illegal – something I am trying to address in this book. The majority of Australian sources amplify the message that feeding leads directly to appalling diseases, dependence on human food (followed inevitably by starvation) and outbreaks of rodents, along with other catastrophes – which probably only adds to the confusion and concern of those of us who are feeding birds. These are all possible outcomes, but they are not a given if we follow certain feeding principles and take a strategic and considered approach.

The dilemma for people who provide food to wild birds is that the sources of these alarming pronouncements include the very organisations a concerned feeder might think of contacting for advice: state and local governments, wildlife carers and veterinarians, for example. It is fairly obvious what these groups would say if you were silly enough to ask them, 'What *should* I be feeding my magpies?'

For all the feeders who are anxious and worried or who even give up feeding altogether, others will conclude that such unfortunate problems don't apply to the situation in *their* yard. These people will probably continue to provide food in such a way as to jeopardise the wellbeing of the birds that visit.

On the other hand, if you persist with the internet sites offering advice on bird feeding, you will soon find examples from *other* countries that offer a very different perspective. From the BBC's *Nature* website ('Providing food for birds is one of the most helpful things you can do for your garden visitors ...') to the Royal Society for the Preservation of Birds' 'Safe Foods for Birds' page, to America's august Audubon Society's 'Bird-Feeding Tips', the message could hardly be more different. Rather than Australia's familiar 'Don't feed!', you find instead 'If you care about birds, feed them'. The contrast is confronting and raises plenty of questions. Feeders in Australia are left wondering why our 'official' attitudes are so negative, when concerns about the effects of feeding on birds are probably similar wherever it is practised.

FOR THE MANY AUSTRALIANS FEEDING WILD BIRDS, THE CHOICE OF COMMERCIAL FOODS AVAILABLE IS EXTREMELY LIMITED.

More practically, however, Australian bird feeders may feel greatly disadvantaged. The volume of information on bird feeding from other parts of the world is enormous. 'How to Feed Birds' books, pamphlets and websites are legion, and much of this information is based on scientific studies, controlled experiments and industry research. This is a truly colossal commercial enterprise, worth billions internationally. Enormous effort goes into quality control and food preference experiments, as well as new product development and marketing. The result is a bewildering array of feeding-related products, including species-specific seed mixes, fat balls, feeder designs and peripherals (such as weather guards and squirrel deterrents). These are widely available in supermarkets, pet stores and specialist birding centres, and from a vast number of online sites.

Here in Australia things are quite a bit different. The range of wild bird feeding products is modest, and those available from the main supermarket chains and pet shops are very often generic and of highly variable quality. A recent examination by ecologist Danielle Hodgson of Australian products found almost all of those labelled 'wild bird' were identical to those developed for cage birds (although the image of a pet cockatiel on the box may have been replaced with an apparently wild lorikeet). Just as common are seed mixes made up of a limited range of cheap cereals with some added sunflower seed and millet,

BUYING WILD BIRD FOOD IN AUSTRALIA

If you are new to feeding wild birds, or maybe rethinking how you go about it, it's important to know where to find appropriate bird food and what to purchase. At the time of writing, the following represent some (definitely not all) of the more relevant products available in Australia. Only those for which most of the ingredients are locally sourced are listed. The companies included here produce high-quality and dependable food products for Australian wild birds, although almost all of the products they sell have been developed for cage birds, or for raising young or injured birds. Bird food specifically designed for wild species is still fairly rare, but specially formulated products are on the way. You can buy these products in stores, order them online, or both.

Please note: This is not an endorsement for the companies and products listed.

SELECTED PRODUCTS	COMPANY
Avione Lorikeets Wild Bird Nectar	Avione Pet Products
Green Valley Grains Wild Bird Mix	Green Valley Grains
Whistler Millet Sprays	Lovitt's Group Pty Ltd
Whistler Avian Science Small Wild Bird Mix	Lovitt's Group Pty Ltd
Whistler Wild Bird Treat (Large and Small)	Lovitt's Group Pty Ltd
Harmony Wild Bird Lorikeet & Honeyeater Mix	Mars Birdcare Australia
Harmony Wild Bird Seed Mix	Mars Birdcare Australia
Harmony No More Waste Seed Mix	Mars Birdcare Australia
Golden Cob Wild Bird Mix	Mars Birdcare Australia
Passwell Complete Lorikeet	Passwell Pty Ltd
Wild Earth Wet & Dry Lorikeet Diet	Vetafarm
Wombaroo Granivore Rearing Mix	Wombaroo Food Products
Wombaroo Insectivore Rearing Mix	Wombaroo Food Products
Passwell Fruit & Nut	Wombaroo Food Products
Lorikeet Delight	Wombaroo Food Products
Wombaroo Lorikeet & Honeyeater Food	Wombaroo Food Products
Wombaroo Lorikeet Nectar	Wombaroo Food Products

sometimes containing 'added natural grit' (in other words, a handful of gravel). As detailed in chapter 5, there are some important exceptions, including carefully developed food products designed with particular wild bird groups in mind (nectarivores or insectivores, for example), but these are much less accessible and are relatively expensive. Indeed, most of these high-quality products are intended for captive or native birds during rehabilitation following injury; they are not meant for attracting wild birds.

At one level, Australia's commercially available wild bird food situation is hardly surprising. The birds that visit feeders here contrast dramatically with those being fed in the northern hemisphere. Visitors to a typical feeder in Europe (blue tits, nuthatches and goldfinches) or North America (chickadees, cardinals or house finches) tend to be smaller grain-eaters (or 'granivores'). Our Aussie regulars are likely to be Australian magpies, rainbow lorikeets or crested pigeons (though rarely at the same time): these are all much bigger, bolder and bossier. And in the case of the most popular bird among Australia's bird feeders, the magpie, the food they have come to enjoy will almost always be some form of meat. Magpies naturally eat invertebrates, extracting worms and grubs from lawns and garden beds, but they have quickly learned to switch to other high-protein items such as beef mince (or 'ground beef') or ham when available. Plenty of other species – kooka-burras, currawongs, butcherbirds and crows – have noticed these offerings and have been quick to take advantage when they have the chance. (Disturbingly, so have rainbow lorikeets but that is another story – see 'Extreme diet switching: seeds and meat' in chapter 6.)

THE BIRDS THAT VISIT FEEDERS HERE CONTRAST DRAMATICALLY WITH THOSE BEING FED IN THE NORTHERN HEMISPHERE. OUR AUSSIE REGULARS ARE MUCH BIGGER, BOLDER AND BOSSIER.

The commercial seed mixes found in supermar-kets and pet shops will certainly be consumed by par-rots, lorikeets, cockatoos and native pigeons, as well as by house sparrows and introduced pigeons. But when it comes to birds with any other type of diet (nectar, small seeds, fruit, berries, insects, for exam-ple), there is very little available from the usual places where people do their shopping. Inevitably, this has led many Australians who feed wild birds to use all sorts of other food items. Often these have been explicitly produced for human consumption, and many are of doubtful nutritional value or are in fact harmful for birds.

A recent investigation found all of the following items being used by feeders in Australia, sometimes in great quantities:

🍁 beef mince	🍁 potato chips
🍁 salami	🍁 cake
🍁 sausage	🍁 donuts
🍁 heart	🍁 sugar
🍁 bread	🍁 honey
🍁 salted peanuts	🍁 cheese
🍁 crisps	🍁 'spoiled kitchen waste'.

Almost nothing is known about the effect of most of these items – including mince and bread, which are used in huge amounts daily (although there has been some study of the consequences of processed meats and chips for bird health).

THE GREAT AUSTRALIAN DENIAL*

So, we have a serious problem. There is a widespread belief in Australia that relatively few people actually feed wild birds. And a strong and persistent attitude that those who do so are generally ill-informed and possibly wilfully ignorant. This perspective was nicely captured by the British ornithologist Andrew Cannon in an article on the importance of private gardens for bird conservation. Andrew asked people around the world for their views on this subject and to his considerable surprise found that: 'Generally, the more conservation-minded and knowledgeable individuals in Australia do not feed their garden birds'.

This is a fascinating quotation for numerous reasons. Firstly, the Australian source of this view was articulating the generally accepted attitude towards bird feeding in this country, one based mainly on a desperately hopeful *perception* rather than on what is really happening. The stark reality is dramatically different. One of the first detailed studies of bird feeding in Brisbane – conducted at the same time as

* Designed mainly to get your attention, this subheading should not be taken too seriously. Indeed, I could provide a long list of contenders for this claim of far greater significance than bird feeding, but it does serve its purpose here.

Andrew Cannon's study – discovered that 38 per cent of households were actively engaged in bird feeding. This news did not go down well, especially among the apparently 'more conservation-minded and knowledgeable individuals'. Although these figures were strongly disputed at the time, numerous other studies conducted over the past few decades have all obtained similar findings: between a third and a half of Australian households are regularly feeding birds. These are rates almost identical to those in other parts of the world, in countries where bird feeding is vigorously promoted by the leading bird and conservation groups.

STUDIES CONDUCTED OVER THE PAST FEW DECADES HAVE FOUND THAT BETWEEN A THIRD AND A HALF OF ALL AUSTRALIAN HOUSEHOLDS ARE REGULARLY FEEDING WILD BIRDS. THESE RATES ARE COMPARABLE WITH THOSE IN COUNTRIES WHERE BIRD FEEDING IS ENCOURAGED BY LEADING BIRD AND CONSERVATION GROUPS.

My attempt to understand the complex global story of wild bird feeding (presented in my book *The*

Birds at My Table, NewSouth Publishing, 2018) confirmed that Australia remains the only country in the world where feeding in gardens is strongly and systematically discouraged. The reasons why this is the case are far from clear. Yes, our birds are different. And most of the country doesn't experience the severe winters typical of much of the northern hemisphere; we simply don't have the 'starving birds in the snow' scenario to prompt a humane response – the situation traditionally used to explain the origins of bird feeding in the United Kingdom, for example. But all the arguments given to justify why people should not be feeding – dependency, disease, inappropriate foods – are concerns shared by feeders around the world. How anti-feeding came to be the predominant attitude in Australia remains a significant mystery. (As is the fact that huge numbers of Aussies feed birds anyway!)

ACTUALLY, WE DO NEED A GUIDE FOR FEEDING WILD BIRDS

I believe that it's time to start a serious conversation about feeding wild birds in Australia. My motivation is, firstly, a genuine concern for the welfare of the birds visiting our feeders. Their health and safety are

AUSTRALIAN FEEDER BIRDS GROUPED BY DIET

Australia has over 700 species of birds, but only a tiny fraction of these are likely to turn up at your feeder. Instead of listing all the possible species, I have referred throughout the book to some general categories based on size and diet. Here is a simple summary, along with examples of some of the birds in each group:

Large granivores
These are the grain- and seed-eaters, so this group includes some of the most common visitors to feeders. Their natural diet features an enormous range of seeds, obtained from the ground and from plants, but these birds may also eat fruit, berries and even insects.
Examples: Galah, king parrot, cockatoos, rosellas, pigeons and doves

Nectarivores (honeyeaters)
A large and diverse group of native nectar- and pollen-eating species, although all these birds also eat a lot of insects.
Examples: Honeyeaters, wattlebirds, Eastern spinebill, silvereyes, noisy miner

Nectarivores (lorikeets)
A fairly small group of nectar (and pollen) specialists, but they also eat a wide range of seeds, fruit and even insects.
Examples: Rainbow lorikeet, scaly-breasted lorikeet

Large insectivores/carnivores
The group, which includes some very familiar birds, normally feeds on all sorts of insects, worms, caterpillars, lizards and baby birds.
Examples: Australian magpie, butcherbirds, laughing kookaburra

Small insectivores
A broad group of species that almost exclusively consume insects. Most are unlikely to visit a feeder, which would typically not offer much that these birds could eat.
Examples: Fairy-wrens, fantails, flycatchers, whistlers

Fructivores
The fruit-eaters, a diverse group with a liking for the juicy and sweet, although these birds also eat plenty of insects.
Examples: Eastern koel, figbird, bowerbirds, some rainforest pigeons

Finches
Although this group should be regulars at feeders, they have not done well in the face of urban development. Maybe appropriate food and backyard habitat could change that.
Examples: Double-barred finch, zebra finch, firetails, mannikins

directly affected by the decisions we make. We have to get this right.

My second motivation is a desire to assist the feeders. My interactions with a large number of people who feed birds throughout the country has convinced me that most are genuinely committed to doing the right thing – but they are frustrated by their inability to obtain useful guidance. Almost all of them care passionately about their wild visitors, yet they continue to feed in the absence of appropriate knowledge, while being unable to seek assistance or advice from the relevant organisations and groups. They know that the response to 'How?' will be always be 'Don't!' We have to get this right, too.

There is another important reason for this book. After a long time spent in the 'Never feed!' camp, I began a slow and careful exploration of the practice, talking with a huge number of feeders, and thinking deeply about the many issues involved. It has led me into unexpected territory. I now believe strongly that the feeding of wild birds can be an important and enjoyable experience, and that practised properly it can be of great benefit to the people involved. People who feed birds demonstrably care, and most know more about their birds, and feel much more connected to nature, than those of us who don't

choose to feed. Whether the birds benefit, however, depends on us. That is largely what this book is about.

The subheading of this book should probably be 'A *tentative* guide'. That is just being honest. We really are a long way from being certain about many aspects of bird feeding in Australia. The reluctance to accept that large numbers of people are feeding enormous numbers of birds daily, throughout the country, has meant that very little direct (and only a small amount of indirect) research of relevance to bird feeding has been conducted here. My sincere hope is that this little book will encourage more scientific studies into all aspects of wild bird feeding in Australia. I would be delighted if *Feeding the Birds at Your Table* was rendered completely irrelevant in the future.

2

THINK

TAKING BIRD FEEDING SERIOUSLY

I suspect that for the majority of people feeding birds, it's just something we do because it's a simple, enjoyable pastime. Having completely wild animals fly in to visit – just by putting out something for them to eat – can be a delightful and satisfying experience. For others, though, providing food is much more than simply a handout. Millions of people who are actively engaged in feeding birds at their homes also care deeply about 'their' birds. The very act of providing food – usually requiring some level of cost in terms of time, money, planning and commitment – demonstrates concern and a desire to be involved. It may be a way of connecting closely with nature or even an attempt at undoing some of the terrible damage humans have inflicted on the natural world. Feeding birds can be a profound experience with strong ethical, philosophical and even spiritual meaning.

But whatever our motivation, if we are going to feed wild birds, there are some things we have to face up to.

This simple, private pastime, which brings delight and joy to so many people, is also an activity that can change entire communities of birds (by helping some species who may drive away others, for example), which in turn can lead to ecological change over large areas. It may even threaten the very existence of the birds we treasure. These are real concerns and they need to be confronted and taken seriously.

FEEDING BIRDS IS NOT NATURAL

Now that is obviously a highly provocative statement, but stay with me. Let's go through my normal morning feeding routine and I will try to explain.

Some rainbow lorikeets and a couple of crested pigeons are perched in the trees above the feeder, waiting for something to happen. My feeder, mounted on the verandah railing, is a round platform — just a big

*flowerpot saucer with some holes drilled through it to let
the rain out. I sweep a few leftover seeds and sunflower
husks (and some possum poo) off and wipe the whole
thing with a vinegar-soaked cloth. I then dry the feeder
thoroughly with another rag, spread a cupful of seed
mix, and retreat back into the house. Through the glass
door I watch a couple of lorikeets and a pigeon land
and immediately jostle for space, a flurry of parries
and beating wings. Eventually, they settle down on
respective sides of the platform, the lorikeets deft and
efficient, the pigeon noisily pounding with its beak. A
king parrot has just turned up, but waits quietly in the
foliage above.*

That's a scene typical at my place and probably
many others all over the country. At least when seed
is offered – if I'd put out meat or bread, the visiting
species would have been very different (see 'Australian
feeder birds grouped by diet' in chapter 1). So why
would I describe this delightful scene as 'unnatural'?
Let's pick it apart carefully.

Firstly, neither of these species would naturally obtain their breakfast in this way, standing on a flat surface, eating seeds that have been deliberately laid out
at their feet. Instead, lorikeets should be scrambling
about through the outer limbs of flowering eucalypts,

callistemons or other native nectar-producing trees, constantly on the move as they search for pollen, nectar, seeds, fruit and even the odd caterpillar. The pigeons, on the other hand, would normally wander about on the ground gathering tiny seeds directly from grasses and small plants. Although both species often gather together in flocks to feed, individuals tend to forage alone, keeping well away from others, concentrating on the space in front of them and ignoring their neighbours. In the wild, they rarely feed close together – and never with other species.

It is also extremely unnatural for birds to congregate in one small place, lots of bodies jostling for space and competing directly for the dense pile of food that has suddenly appeared. Nor is it normal for there to be aggression and sometimes even violent interactions between the birds, especially between species of different sizes. Most birds would avoid such confrontations and simply move away. However, this pile of large and nutritious seeds is apparently worth scrapping over.

Just about all of this is unusual: dense clusters of birds aggressively competing for a pile of food that arrives at the same place and same time virtually every day. Such predictability is just not normal or natural.

And what is it that they are eating? My seed mix is similar to most commercial products, made up of various exotic seeds originating from places all around the world: sunflowers were first cultivated in North America, wheat in the Middle East, millet in India, sorghum in northern Africa. All of these are grown in Australia these days, of course, but none were a part of the natural pre-European era diet of these species, or indeed of any of the birds visiting feeders throughout the country. Almost everything we place on or in our feeders is 'unnatural' in this sense, and that goes for the seeds consumed in colossal quantities by birds on every continent.

(I say 'almost' because there are some people who actually gather native seeds – such as wattles, eucalypts and numerous native grasses – to feed their birds, and there are even some commercial mixes that include various indigenous plant seeds.)

Apart from a very small number of cases – the few granivorous species that consume the tiny seeds of the indigenous sunflowers of Mexico and Arizona, for example – hardly any birds visiting feeders anywhere in the world are consuming items that featured in their ancestors' natural diet.

And so far we have only been talking about seed. When we think about the human foods on offer (bread, mince, cheese, salami, chips, biscuits, Belgian dark chocolate with added Mediterranean salt ...), we start to see how far from 'natural' things can get.

BIRDS MAY CHOOSE TO EAT ONLY PARTICULAR SEEDS (OR NONE) ON THE FEEDER THEY VISIT. MANY SPECIES CAN DETECT COMPONENTS SUCH AS OILS, PROTEIN AND EVEN MINUTE AMOUNTS OF CERTAIN AMINO ACIDS, AND WILL SELECT THEIR FOOD ACCORDING TO THEIR NEEDS, DEPENDING ON THE SEASON AND THE STAGE OF THEIR BREEDING CYCLE.

The historical origin of the seeds is, however, not really that big a deal. All of these familiar seed types have undergone millennia of systematic breeding to greatly increase their size, palatability and nutritional quality, in most cases for human consumption. In fact, these cultivars are greatly superior to their ancient progenitors in terms of value as a food. Any cockatiel, rosella or finch would sensibly select the modern version over the original if they had the choice. Although there are plenty of exceptions, most commercial seed mixes do provide a half-decent snack, although usually not the full complement of nutrients that birds require. And even though some of the generic products available in the shops contain large proportions of cheap, hard cereals (such as oats, wheat and sorghum), the birds can and do simply ignore the junk. We now know that many species are able to detect particular components such as oils, protein and even minuscule amounts of certain amino acids in the foods they select. Sometimes they need more protein, sometimes more calcium, depending on the season and the stage of their breeding cycle (see 'What birds need to eat' in chapter 3). As a result, there will be times when our expensive Wild Bird Supreme with MegaVita Plus might be completely ignored as the birds binge on some naturally occurring seed available in the bush nearby. This ability of birds to choose particular seeds on the feeder, or to ignore the lot, is completely normal.

It is important to appreciate that most of the birds visiting our feeders will obtain most of their daily sustenance elsewhere. As will become clear, the birds visiting our feeders usually don't *need* to come. They get most of what they require the natural way. But a snack and catch-up can also be nice (just like it is for us).

The point of this discussion is not to emphasise the unnatural character of what we are doing when we feed birds, but rather to start us thinking about some of the wider issues. When we offer food to birds we change things. Whether these things matter will depend on our attitude, level of knowledge and the extent to which we take responsibility for our role.

THE ISSUES WE NEED TO UNDERSTAND

Everyone with even a mild interest in wild bird feeding in Australia will be aware of a range of issues that are offered as arguments against feeding. Many of these are also regarded as serious by the very people who regularly feed birds.

The problem is that those opposed to feeding accuse the feeders of ignoring the facts, while the feeders believe that the negative claims are exaggerated by people who simply don't understand. This standoff has been around for decades and has changed nothing: haters gonna hate, while the feeders still keep on feedin'.

In response to this dilemma, I want to provide a summary of what is known about the main issues. I have based my conclusions on as many published research results, official reports and scientific studies as I could find, although the information available is still fairly fragmentary. In addition, I have discussed these topics with a wide range of veterinarians, wild-life ecologists, bird nutritionists, zoo personnel and other specialists.

The most obvious and important issues are listed below, ordered roughly by the amount of attention they receive from researchers, commentators and online sources. Importantly, they have all been iden-tified – by several studies – as being of significant concern to both opponents and supporters of wild bird feeding.

1. Dependency (Do birds become reliant on the food we provide?)
2. Disease (Does feeding spread disease?)
3. Nutrition (Is the food we provide suitable and nutritious?)
4. Dominance (Does feeding change the interactions among local birds?)
5. Interlopers (Do feeders attract predators and vermin?)
6. Conflict (Does feeding lead to problems with (human) neighbours?)

Below, I explore what is known with certainty and what is still a matter of conjecture. The amount of detail is based on how much information is avail-able and how much we need to think about the issue. Some apparently critical concerns turned out to be a lot less serious than I at first thought. Others are much more significant than I expected. I also make some simple suggestions for what practical steps can be taken to mitigate, minimise or eliminate each of these concerns (more detailed information about the most significant issues can be found in later chapters).

To make sure the messages are not lost in the detail, I have provided a simple 1 to 5 🐦 rating of the *significance* of the issue for feeders in Australia. If it's really important, we need to be ready to act.

DEPENDENCY

SIGNIFICANCE: 🐦🐦🐦🐦🐦

This seems to be the big one. In numerous studies undertaken around the world, the topic of dependency is the first to be mentioned by feeders and anti-feeders alike. It is the primary reason for the so-called Golden Rule of Bird Feeding: once you start, you must not stop. Because if you do, the birds may forget how to forage or what their natural foods are, leading to hunger and loss of condition; they may even, in the worst-case scenario, starve to death. These are powerful concerns, and they have been responsible for

many people feeling they must provide an uninter-
rupted supply of food at all times. I know of plenty
of people who will not leave their homes for more
than a few days because of this – and extended trips
or holidays are simply out of the question. Having
what looks like the same birds continually visiting the
feeder has led these people to conclude that 'the birds
must always be hungry'. Clearly it would be unethical
to fail to maintain the supply if this were true.

Cynically, it would be tempting to suggest that the
Golden Rule was invented, or at least maintained, by
the bird food industry in order to keep sales up. There
is, however, some basis to the idea that birds may
become dependent on a steady supply of human-
provided food, although usually in an extremely
limited way. At certain times and in certain places,
feeders (and from this point on I am usually refer-
ring to the physical apparatus used to supply food) do
become the main – or only – supply of seed. This can
occur during extreme weather events such as severe
and extended snowstorms, as was the case throughout
parts of the United Kingdom during the winter of
2017–18. A series of brutal storms (dubbed the 'beast
from the east') dumped massive amounts of snow and
devastated woodlands and hedgerows. The destruction of much of their usual winter food supplies

resulted in astonishing numbers of birds, including species almost never seen in urban areas, descending on private gardens in search of something to eat. The British media exhorted people to refill their feeders or provide anything that the starving and distressed birds might be able to eat. In this case, the birds were genuinely dependent on the feeders. But only temporarily. When the cold retreated, so did the birds, back to the countryside. The twitchers of London had to be content with the usual blue tits and goldfinches, rather than desperate flocks of fieldfares and redpolls.

This event was just one particularly dramatic version of the usual winter scenario typical of much of the northern hemisphere. The tradition of providing for the birds that do not migrate away with the start of the cold weather is the most familiar form of bird feeding in the world. Many Australians with European ancestry can recall grandma's suet balls or peanut cakes, concoctions laden with fat and protein that attracted swarms of hungry birds back in the old country. Although there is an increasing tendency to feed year-round in the United Kingdom and Germany, for most of northern Europe bird feeding remains a decidedly winter-only activity – feeding when the weather is mild is usually severely frowned upon. The reason normally given for discouraging feeding in other seasons is 'to make sure the birds don't get used to handouts and are able to fend for themselves when winter is over'. In other words, these traditional restrictions are aimed at minimising the possibility of dependency.

THE BIRDS VISITING OUR FEEDERS USUALLY DON'T *NEED* TO COME. THE MAJORITY WILL OBTAIN MOST OF THEIR DAILY SUSTENANCE ELSEWHERE, GETTING THE BULK OF WHAT THEY NEED THE NATURAL WAY.

There are some other important examples of birds becoming heavily reliant on feeders. These all relate to one of the most serious unintended consequences of bird feeding: the failure to migrate. In northern Scandinavia, most small birds migrate south with the start of winter – a sensible behaviour, as the cold in this region can be severe. During the 1980s, however, it was noted that some of the normally migratory great tits were failing to leave and had become entirely dependent on garden feeders – for many months, this was literally the only food available to them. Similar, but somehow much more disturbing, is the case of the gorgeous and delicate

Anna's hummingbird, which has long been a temporary summer visitor as far north as Seattle and Vancouver in the west of North America. When it turns cold, these tiny birds (they weigh just a few grams) traditionally head south to California. People really love these birds, and there has been an explosion in the number of hummingbird feeders (filled with the sugar-water that these birds crave) provided in the northern areas; these include special feeders designed for winter conditions, with inbuilt heaters to prevent the sugar-water from freezing. This has resulted in many of these sensitive birds remaining in the extreme northern parts of their range throughout the winter, utterly and terrifyingly dependent on these artificial sources of 'food'. Should anything go wrong, the birds are almost certainly going to perish.

These rare but real examples of genuine dependency are actually the exceptions that prove the rule. The reality is that despite the enormous number of bird feeders that are visited daily, birds will obtain most of their food from the natural environment. Even in large towns and cities, where it would appear that the easiest way for birds to forage would be to concentrate on feeders, almost all birds still find and eat a largely natural diet.

Some recent findings involving Australia's most commonly fed bird, the magpie, have particular relevance to the issue of dependency. Given the popularity and abundance of this iconic bird in urban areas, and the ubiquity of magpie feeders, it would appear quite possible for the species to give up feeding naturally altogether. After all, trays of mince, cheese, ham and diced heart are available in gardens in every second street. Surely the temptation to switch to an all-human diet would be particularly strong for exhausted parent magpies attempting to satisfy their ravenous nestlings: they could simply start ferrying mince and cheese from the feeders to their nests, and give up searching for worms for their chicks.

This was exactly our expectation when my students and I studied magpies in Brisbane. Yet to our considerable surprise, less than a tenth of the food brought to the chicks was obtained from feeders. While it would definitely have been easier and quicker to take advantage of all that mince on the feeders, we discovered that most magpie parents have a 'no junk food' policy when feeding their chicks, insisting on worms and grubs procured the old-fashioned way.

This somewhat unexpected finding for Australian magpies turned out to be quite typical of a wide variety of bird species in similar situations all around the

world. Even though birds may visit feeders regularly, numerous studies have found that most of their daily diet is natural. The actual proportion varies enormously across species, season and locations, but on average most birds obtain much less than a quarter of their daily diet from feeders. Not unexpectedly, this proportion tends to be higher in colder weather, while in late summer it is common for feeders to be abandoned altogether, regardless of how well supplied the feeder is.

This discovery is actually an enormous relief – in fact, it may be one of the most important messages to emerge from all the research done on bird feeding. Most of the time, birds really do not become dependent on the food we provide. Rather than 'needing' our feeders, most species on most days regard the food supplied by humans as a welcome but far from essential supplement.

This revelation may also be something of a disappointment, given that all the effort and care (and expense) we put into buying the food, cleaning the feeder and providing a reliable food source may not have as much significance as we'd imagined. Although this may be an unwelcome reality, in general, feeding birds is more important for us than it is for the birds. We feed them for a complex array of reasons and obtain enormous enjoyment and satisfaction from the experience. It can be calming, exciting, healing, connecting and illuminating. These brief encounters with truly wild creatures can be priceless for us. For the birds, it is probably just a snack.

WHAT TO DO

If we can accept the fact that most of the time the birds don't actually need the food we provide for them, and that their visits make up just a tiny part of their busy lives, we should be able to relax a little. Missing a day of feeding or going on a trip will not be a serious problem.

The best approach is to provide only modest amounts of food – just a cupful or a handful is plenty. Of course, the birds will finish it off pretty quickly, but that is fine. You can also vary what you put out. Birds are always on the lookout for new opportunities, and mixing it up on the feeder is a good idea: seed mix one day, grapes and blueberries the next; for the carnivores, some pet food followed by mealworms (you'll find lots of suggestions in chapter 4). Watching the birds' reactions to unexpected items on the feeder can be a lot of fun.

DEPENDENCY IN SUMMARY

- Most birds visiting feeders obtain most of their food from natural sources.

- There is very little evidence that birds become dependent on feeders for their daily sustenance. The main exceptions are found when reliance develops during extreme weather events, or when some birds no longer migrate as a result of being fed.

- People are more likely to be addicted to the practice of feeding than the birds are!

- To discourage any chance of dependency, only provide small amounts of food and supply a variety of types.

DISEASE

SIGNIFICANCE: ✿✿✿✿✿

If dependency is an apparently serious issue that probably *isn't*, disease is the opposite. No other issue presents such an acute threat to the fundamental aspects of the practice of feeding wild birds: the enjoyment, the benefits, the ecology, the wellbeing of the birds themselves. If that sounds a bit dramatic, it is meant to. This element of the bird feeding story must be taken very seriously indeed. The issue of disease has been neglected – even ignored – and generally minimised, leading to apathy and ignorance among the feeding community. Just as we need to examine the issue of dependency in a critical and focused way, the relationship between our provision of food for wild birds and the potential spread of disease needs careful scrutiny.

One of the very first problems when talking about this issue is the lack of direct evidence: most of the time you just don't see sick or dead birds. However, two of the most catastrophic epidemics associated with feeders – house finch disease in the United States and finch trichomoniasis in the United Kingdom – were first noticed because sick birds often

40

remained at the feeders. Both diseases impaired the birds' vision and led to 'extreme lethargy', which meant they stayed close to the easiest source of food. Dead or dying birds were also found next to feeders. These diseases each had a massive impact on the populations of affected birds, in the case of the British greenfinch reducing numbers by more than a third in a single year.

IF WE ARE GOING TO PROVIDE FOOD FOR WILD BIRDS, MINIMISING THE POSSIBILITY OF SPREADING DISEASE HAS TO BE OUR GREATEST PRIORITY. THIS MUST INVOLVE THOROUGH, DAILY CLEANING OF ALL SURFACES WHERE FOOD IS PLACED AND WHERE BIRDS STAND AND FEED.

Major outbreaks in other countries are mentioned here because they are an extreme exception to the normal situation. As I mentioned above, it's very unusual for people to see dead wild birds, which are much more likely to die unseen, away from feeders. The bodies of most birds are fairly delicate and light and simply don't last for long; they decompose quickly or are consumed by predators. This means that we are unlikely to see any evidence of a disease outbreak unless it is overwhelmingly severe, with birds dying in numbers significant enough to draw our attention.

As far as is known, there have not been any major epidemics among the birds that visit feeders in Australia. On the face of it, that is good news, but it may also be because there are simply fewer people watching. We know a great deal about the two overseas epidemics because huge numbers of people were already actively engaged in recording the birds that were visiting their feeders, through their participation in two major citizen science programs: *Project Feeder-Watch* in North America and *Garden BirdWatch* in Britain. Both have tens of thousands of loyal members, who routinely send in their sightings every month. When the first indications of each disease began to be reported, these organisations were able to very rapidly alert their participants to be on the lookout. This enabled the geography, spread and longevity of these outbreaks to be followed, mapped and investigated in extraordinary detail. (Australia does have a similar program, *Birds in Backyards*, run by BirdLife Australia, but this is most definitely not focused on feeders.)

There is a tragic irony in having such detailed knowledge about these terrible diseases because of the diligence of the people watching the birds at their feeders. Because these epidemics were almost certainly *due* to the feeders. The organisms responsible for house finch disease and finch trichomoniasis persist very successfully among the seeds and droppings left at feeders. These organisms are also spread during direct bird-to-bird interactions. By feeding, we are intentionally concentrating lots of birds, and often different species, together in a small location, where direct contact is frequent. It is absolutely no surprise that such a situation is ideal for the spread of infections.

WHEN WE FEED BIRDS WE ARE INTENTIONALLY CONCENTRATING LOTS OF BIRDS, AND OFTEN DIFFERENT SPECIES, TOGETHER IN A SMALL LOCATION – CREATING AN IDEAL SITUATION FOR THE SPREAD OF INFECTIONS.

Australia may not yet have experienced an epidemic due to the presence of feeders, but the veterinary specialists I contacted say such an event is almost inevitable. Many of the pathogens known to cause a wide range of bird diseases are here already, ubiquitous but at naturally low levels in wild populations, as well as being present in domestic poultry, cage birds and feral pigeons.

A potential candidate for a significant outbreak in Australia is beak and feather disease, often referred to simply as 'bald cocky disease', which affects many of our parrot species. The pathogen, which has been traced to a virus, is known by the technical label psittacine circoviral disease. It results in alarming feather loss, distortion and discolouration, and sometimes in deformation of the beak. Birds that contract the disease typically do not respond to any treatment, and will ultimately die. Beak and feather disease occurs naturally among almost all of Australia's parrots, lorikeets and cockatoos, and it appears to have been present here well before people arrived. Outbreaks in natural populations have probably occurred irregularly, leading to periodic crashes and recovery. In the 'big picture', natural scheme of things, most species can probably cope with these fluctuations in population, provided they are widely distributed and relatively abundant. It is the species with low numbers and smaller ranges that are much more vulnerable, obviously.

This is exactly the reason that many agencies and experts are particularly worried about beak and feather disease. Although there have been several major outbreaks among wild galahs and sulphur-crested cockatoos, possible symptoms have been detected among rainbow lorikeets from urban areas in numerous locations on the east coast. Although the connection with feeders has not yet been proven, many veterinarians have pointed out the very obvious fact that rainbows are among the most frequent visitors to feeders throughout the country, and the prospect of this disease spreading rapidly from bird to bird via feeders is all too plausible. Add to that the reality that lots of other parrots also visit these same feeders, and the possible results are genuinely alarming. This is a scenario we must avoid at all costs.

WHAT TO DO

The data may be far from clear in Australia, but the experience in many other parts of the world could not be more serious or direct: feeders do assist in the rapid spread of diseases. By their very design, platform feeders bring many birds of numerous species together to feed on a flat surface with a limited area – where food and faeces can freely mix. It's a situation asking for trouble. The risk is significantly less for hanging vertical feeders, but these are also strongly implicated in enabling infection (almost all of the feeders used by the house finches and greenfinches were of this design).

There is simply no alternative: if we are going to provide food for wild birds, minimising the possibility of spreading infection must be our greatest priority. This must involve regular – daily – and thorough cleaning of all surfaces where you place the birds' food and where birds stand and feed (see '*Really* cleaning your feeder' in chapter 3). This cleaning process usually involves three key steps:

- sweeping
- cleansing
- drying.

Sweeping involves removing all leftovers and waste from the feeding surface, using a small brush (which must also be cleaned carefully and regularly). Preferably, you should collect and dispose of this debris, especially if it is heavily soiled with droppings. Simply sweeping the stuff onto the ground below may allow potentially infectious material to be consumed later. Best to bag and stick in the rubbish.

Cleansing is the crucial step. You must make a serious attempt to eradicate the organisms responsible for bird diseases from the feeding surface – or at least reduce the possibility that they are present. There are plenty of suitable cleaning products around but a feeder is, after all, a place for eating, so we shouldn't use anything toxic, such as the ammonia-based detergents (see '*Really* cleaning your feeder').

Drying is essential to ensure that no cleaning fluids remain on the surface in liquid form. It is also important to have an entirely dry surface in preparation for feeding: almost all of the organisms we are concerned about survive in moist environments. Having a completely dry surface makes this much more difficult.

These directions are most relevant for the typical flat feeding stations (platforms, hanging 'house' feeders and traditional bird tables). Hanging tube feeders and hopper feeders are less risky because droppings don't usually accumulate on the feeding surface, and fewer birds can congregate closely together. But the area *beneath* the feeder is another matter altogether. The accumulation of faeces and leftovers in a damp location, often over extended periods, makes such a place particularly dangerous for any birds foraging there. These concerns are obviously relevant for species that feed on the ground, such as pigeons and finches. It is extremely important that these spots be cleaned up frequently to minimise the build-up of nasties.

DISEASE IN SUMMARY

🦅 Feeders bring lots of birds of different species together, to feed in the same small space. This is the ideal environment for spreading infection.

🦅 Mixing bird food, manure, moisture and mouths is dangerous!

🦅 Preventing and minimising the possibility of spreading disease must be our top priority.

🦅 It is essential to clean the feeder regularly by sweeping, cleansing and drying.

NUTRITION

SIGNIFICANCE: ★★★★☆

By 'nutrition', I mean the suitability of the food items being offered to birds as well as their nutritional attributes. A discussion of bird food inevitably draws attention to one of the great differences between Australia and the countries where feeding is much more visible and accepted. The global bird food industry, which is almost entirely based in the northern hemisphere, is staggeringly huge in terms of the amount of seed that is sold and purchased, and the variety of feeders and related paraphernalia that are available. While cheap, dodgy products of dubious value are found everywhere, the commercial bird food products available in Europe and North America are, overall, of excellent quality. Consumer demand for the highest possible quality bird food – overwhelmingly in the form of seed mixes – has led to intense competition among manufacturers vying to offer products of a superior standard. Vast amounts of research have resulted in continuous improvements in the quality of nutritional combinations and vitamin and micro-mineral additives. This is as much about business incentive (to sell more products) as it is about concern for the welfare of the birds: consumers are demanding the best possible quality and are willing to pay for it.

IF YOU HAVE MAINLY BIGGER BIRDS VISITING, BUT WANT TO ATTRACT SMALLER SPECIES, YOU COULD USE COMPLETELY SEPARATE FEEDERS: PERHAPS A SMALLER HANGING PLATFORM OR A TUBE FEEDER, AS THESE ARE MUCH HARDER FOR LARGE BIRDS TO PERCH ON. SET THEM UP WELL AWAY FROM THE 'BIG BIRD' TABLE AND FILL WITH AN APPROPRIATE MIX.

In Britain, for example, bird food companies seek the endorsement of the Birdcare Standards Association, which has determined optimal nutritional combinations of the ingredients used in seed mixes. If approved, the products may display the Birdcare logo, 'proving' that the mix conforms to an independently determined standard. The companies competing for the attention of the bird feeding public maintain a fierce rivalry, leading to some remarkable (and often dubious) claims. A recent manifestation of this competition is the targeting of specific species known to be favourites among feeders. There is now an endless pipeline of new and steadily 'improved' products claiming to attract, for example, goldfinches, cardinals or blue tits. These and an almost unbelievable array of other products are available in a wide variety of outlets and online. The average British, Canadian or American feeding enthusiast may be bewildered by the number of choices but, by and large, they can be certain that the food will be good for the birds.

The contrast with Australia could hardly be greater. Although there are quite a lot of products apparently intended for 'wild birds' (as opposed to cage birds) available in supermarkets, pet shops and online, only a few of these have been specifically formulated with native Australian species in mind. As the primary targets for these seed mixes are the various parrot-like species (rosellas, lorikeets and cockatoos), the products for sale are typically generic combinations of the usual seed types. Although the product may claim to be 'specially prepared for native species', it is often mainly composed of an entirely predictable collection of sunflower seeds, cheap cereals and millet. This may not really be a significant issue – the birds eat most seeds and generally these seem to be of a high standard – but there is currently an extremely limited variety available. This is another direct result of the reluctance to discuss bird feeding in Australia. The manufacturers are understandably wary of spending very much on research and development locally, when political and community attitudes are so disparate and uncertain. As things currently stand, the big pet food producers seem to be waiting to see whether the opposition to feeding is diminishing. If this happens, expect to see a plethora of completely new, specialised products quickly appear on the shelves. If Australia were to follow the northern hemisphere trend, look for packages designed specifically for, say, finches, honeyeaters, fairy-wrens, drongos, robins, bowerbirds or silvereyes. Not only that, you might even see all of these products being used in a variety of feeders in the same garden, if it happened to be in a suitable location.

That may sound extravagant and unrealistic. But it is the equivalent of the daily experience of people who feed birds in Arizona, Poland or Scotland, for example. I recall one snowy backyard in rural New York State, where I counted 12 different species visiting the numerous feeders. Or the tiny garden of a semi-detached home in London where over a few hours blue tits, great tits, nuthatches, ring-necked parakeets, house sparrows, goldcrests, blackbirds, robins and song thrushes all paused briefly to snack at the bird tables, suet balls and hopper feeders. The secret to such a remarkable diversity of species was, of course, a corresponding diversity of foods. In both of these examples, the owners had installed numerous devices that supplied a wide range of high-quality bird food products, all formulated to attract and provide suitable nutrition for different birds.

In Australia, by contrast, the top three foods offered to birds are, in order:

- seed mix
- various types of meat
- bread.

The seed mixes are almost always a standard combination of seeds known to be consumed by caged parrots. The meat and bread, obviously, are simply household items produced for humans. The types of birds these types of food are going to attract are all too predictable.

SEEDS

We have already discussed seed mixes a little; the main point is not necessarily the quality of the components but, rather, which species will consume them. In Australia, seeds attract (unsurprisingly) granivores

– seed-eaters – and these will almost always be the local parrot-type species: rosellas, lorikeets, cockatoos, sometimes king parrots and various pigeons. These are all fairly large birds, and competition can be intense. The chances of winning tend to be based directly on size, although personality is undoubtedly important – rainbow lorikeets often see off species significantly larger than themselves. And of course, smaller or more timid species have little show. This is especially the case with tiny species such as finches and mannikins – and pale-headed rosellas, the smaller lorikeets and most doves also come to mind. (The issue of dominance at feeders is discussed in more detail below.)

PURCHASE SEED PRODUCTS MANUFACTURED BY MAJOR PET FOOD AND LIVESTOCK COMPANIES OR SPECIALIST WILD ANIMAL FOOD COMPANIES – THESE COMPANIES ARE COMMITTED TO ENSURING THEIR INGREDIENTS ARE OF THE HIGHEST QUALITY.

This raises one of the fundamentals of bird feeding: feeders only attract 'feeder' birds. In other words, you get what the menu dictates. If all you seek are big, noisy seed-hogs, stick out the cheap seed. But if you want the smaller species (assuming they are present in your location) you might have to try other approaches. This could include completely separate feeders, perhaps a smaller hanging platform or a tube feeder, which suit the little guys and are actually much harder for the big ones to perch on (see chapter 3). These could be set up well away from the main 'big bird' table and filled with an appropriate mix.

Although wild birds will probably eat the cheap mixes, it is usually a good idea to purchase seed products manufactured by the main pet food and livestock companies or the specialist wild animal food companies (see 'Buying wild bird food in Australia' in chapter 1). These companies are committed to ensuring their ingredients are of the highest quality. They obtain most of their seeds through organisations such as Queensland Agricultural Merchants, who in turn require that their suppliers meet stringent quality standards; these requirements are published as the Australian Bird Seed Standards. There is no such quality control applied to the generic – and cheaper – seed mixes widely available in supermarkets, which typically originate overseas. Best to stick with the known pet food brands that buy local produce for their mixes. (I give more information on seeds as bird food in chapter 4.)

MEAT

If there is one thing that distinguishes wild bird feeding in Australia from the rest of the world it is this: meat being provided instead of birdseed. The only exception is an extremely localised example from Reading in England, where people feed meat to a large bird of prey, the red kite, in their backyards.

The reason for all this meat is pretty obvious. Magpies. The Australian magpie is not only our favourite bird, it is also the most welcome and the most commonly fed; people who feed birds want to attract magpies more than any other species. It's not hard to understand why. Magpies are abundant in towns and cities throughout Australia. As we all know, some of these birds may turn against some of these people, swooping and attacking humans who approach their nests during the breeding season (that's another story), but not those who feed them. These scary marauders can be tamed almost instantly with a handful of mince – a well-known strategy, even among many staunch anti-feeders.

Yes, mince, the virtually ubiquitous food supplied to magpies every day throughout this land. Mince (or 'ground beef'), a cheap, all-purpose meat product, is easily converted into a simple family meal and is therefore omnipresent in the fridges of Australia.

I once attempted to calculate the amount of mince being consumed by magpies in a typical Australian suburb every day and was so shocked that I gave up. But it is a lot. And that is a very serious problem. Because mince is terrible bird food.

This is probably the biggest challenge I will present in this book: we have to give up feeding mince to magpies. It is going to be hard, but it is vital. This simple, cheap, convenient food is a serious threat to the health and wellbeing of our favourite species – and all the butcherbirds, currawongs, kookaburras, crows, ravens, koels and even lorikeets that also love the stuff (see 'Extreme diet switching: seeds and meat' in chapter 6).

THIS IS PROBABLY THE BIGGEST CHALLENGE
PRESENTED IN THIS BOOK: WE HAVE TO
GIVE UP FEEDING MINCE TO MAGPIES.

The problem with mince is twofold. Firstly, it is an incomplete food. Typically, when natural prey is eaten, much more than just the flesh is ingested: feathers, fur, tendons, insect exoskeleton, even bits of bone. When all this is consumed, the ratio of calcium to phosphorus is about even, allowing the predator to assimilate the calcium into its body for a range of purposes. This vital metabolic process, which enables bone repair, eggshell formation and many other tasks, is seriously impaired if the ratio is uneven.

The other really serious issue associated with mince is the relationship between poor nutrition and parasites. These might seem to be completely different topics, but we now know that they are closely connected. The number one problem parasite for magpies, according to vets, is known as 'gapeworm' (see 'Mince, metabolic bone disease and gapeworm' in this chapter).

A less obvious problem with mince is its weird consistency. The mashing and grinding involved in producing this form of meat makes it strangely sticky and lumpy. This is great for stir-fry and spaghetti bolognese, but not for birds attempting to swallow the stuff. Portions can get caught on the concave interior of the beak, and these appear to be very difficult to dislodge. If the food remains in the beak, bacterial infestations can occur, sometimes leading to infections.

It is definitely time to forget about straight mince as bird food. So, what *are* the food options available for meat-eaters? I have had a lot of conversations with wildlife veterinarians about this issue and it is

MINCE, METABOLIC BONE DISEASE AND GAPEWORM

There are some serious problems with feeding mince to magpies, butcherbirds, kookaburras and other insectivorous or carnivorous birds. These problems are only likely to become serious if a bird's diet is made up of too much of the stuff. A little eaten every now and then is not an issue, but some individuals may consume lots fairly regularly and that is when the medical concerns arise.

When any of those species mentioned above eat their natural prey – worms, beetle larvae, skinks, sometimes baby birds and small mammals – they normally consume the entire body: flesh, blood, feathers, tiny bones, dirt, intestines, everything. As a result, they acquire a wide range of vitamins and minerals, including those required for normal metabolism. When birds eat mince as a significant part of their diet, they receive only a fraction of certain nutritional components that they need, with calcium and vitamin D3 in particular being in low supply. By contrast, the proportion of phosphorus in mince is high; without the fur and bones the calcium:phosphorus ratio is about 1:17. If too much mince is consumed, the animal may be unable to absorb the calcium in its diet and start to extract this element from its own bones, which can ultimately lead to metabolic bone disease.

Metabolic bone disease

Metabolic bone disease refers to the suite of conditions in animals where the bones become weakened or malformed, mainly because of too little calcium. Calcium is critically important for the continual strengthening of the skeleton, for muscle contraction and for many other important functions. If there is not enough calcium circulating in the animal's blood, the body will extract this essential building block from its main storage site, the bones, causing them to weaken and become brittle.

In adult birds, metabolic bone disease can result in problems such as rickets, tremors and lethargy. In young birds, the results can be even worse, because their skeleton is still forming. Because nestlings are completely dependent on the food their parents bring them, they are especially susceptible.

If the adults rely too much on mince from the nearby feeder and not enough on worms and grubs, there can be big trouble. Wildlife vets have to deal with young magpies with horribly misshapen legs, distorted beaks, fragile bones and very little energy, which may have been caused by too much mince.

Gapeworm and other parasites

Because mince is such a low-quality food, birds that consume too much, too frequently are likely to be in a poor physical state (and nestlings are obviously particularly vulnerable). This makes them much more susceptible to any of the usual illnesses and problems that healthy birds can overcome, including parasites.

All animals naturally carry a wide range of parasites inside and outside their bodies. Most of the time, these have little impact. But if an individual is weak and malnourished, big problems can occur.

Gapeworm is just one of many parasites that affect Australian feeder birds, but it gets special mention here because it is so commonly found in magpies (and sometimes butcherbirds) that have consumed too much mince. The birds may already be sick, weak or old, and therefore relying too much on mince instead of obtaining a mainly natural diet.

These nematodes are found in the windpipe and normally don't cause too much trouble. However, when gapeworms build up in numbers they can seriously disrupt the ability of the birds to breath (hence 'gaping'). It is not entirely clear why, but birds in a poor nutritional state are extremely prone to heavy loads of gapeworm.

All of the wildlife vets I spoke to implored me to make this point strongly: mince is not suitable food for any birds. Gapeworm is just one reason to avoid it.

clear that there is still a long way to go before we have an ideal alternative. You can choose to keep feeding mince if you want – although this is not the ideal – but you will need to counteract the lack of calcium by adding digestible calcium powder or a commercially available insectivore supplement. Until there are magpie-specific commercial products available, the following foods (discussed further in chapter 4), seem to be the safest options:

- commercial insectivore food
- homemade 'dry porridge'
- chopped meat roll pet food ('dog sausage') with supplement
- chunky tinned dog food
- dry puppy kibble, soaked in water
- garden worms
- mealworms.

BREAD

Yes, bread. Although primarily associated with the feeding of waterfowl (in excessive amounts) at ponds in public places, bread is still one of the most common food types placed on feeders in private backyards. Like mince, it is a popular bird food because it is easy to obtain and cheap, and every home has plenty of it to hand. But – like mince – it is terrible bird food. There are many reasons why we should give up feeding bread to birds, but changing this practice, especially in the context of feeding the ducks down at the park, is going to be a serious challenge.

Bread itself is not harmful to birds, and small amounts are unlikely to have much effect on wild birds that consume an otherwise mainly natural diet. A little offered on your feeder at home is also not a big problem. It is, however, of minimal value to the bird's diet, providing hardly any nutrients while tending to act as filler, potentially leaving less 'space' for better foods. When bread – primarily a source of carbohydrate – dominates the diet of a bird, numerous metabolic problems can arise. Perhaps the most alarming outcome is the horrible and frequently fatal condition known as 'angel wing', which is discussed in 'Bread and angel wing' in this chapter. This

---

A LITTLE BREAD OFFERED ON YOUR FEEDER AT HOME IS NOT A BIG PROBLEM. IT IS, HOWEVER, OF MINIMAL VALUE TO THE BIRD'S DIET, LEAVING LESS 'SPACE' FOR BETTER FOODS.

---

condition doesn't appear to be widespread in Australia yet, but the associated conditions – lots of bread being fed to lots of waterbirds in urban ponds – are all in place. It is possible to feed the ducks safely, but leave the bread at home (see 'What can we feed the ducks?' in chapter 4).

WHAT TO DO

If you are attracting seed-eaters, select products suitable for wild species and make sure that they are manufactured by companies associated with pet or animal food production. In addition to the small number of huge multinationals, there are now several excellent Australian companies producing bird mixes of a high quality. Avoid the generic products found in most supermarkets.

Give up thinking that mince and bread are suitable food for wild birds. Just because they eat it readily, does not mean they should. If we are going to continue to feed magpies, butcherbirds, kookaburras and the like, we need to commit to finding a suitable 'meaty' alternative to mince. Check out the various insectivore supplements that can be added to the food you offer.

NUTRITION IN SUMMARY

🌿 The predominant anti-feeding attitude in Australia has resulted in very little research being done on suitable foods and a limited variety of commercial products.

🌿 Most seed mixes are of variable quality. Always buy products from pet food companies.

🌿 Magpies are our favourite feeder birds and most are fed mince. Don't do this; use pet food instead.

🌿 While not harmful in small quantities, bread is not a suitable food for birds.

DOMINANCE

SIGNIFICANCE: 🍃🍃🍃🍃🍃

When we offer the birds that live around us something to eat, we are changing all sorts of things about their environment and behaviour. That has always been the case. Whether it be kitchen scraps on the back lawn, Premium Wildbird WonderMix in the Acme Supreme feeder, or even the profusely flowering Robyn Gordon grevillea that we planted in the front garden, we are altering the supply of local resources and encouraging the birds to do something they would not normally do. What does *not* change is the birds' natural propensity to continuously look for foraging opportunities – or even experiment with types of food they may not be used to. When the current supply of whatever they are feeding on is finished, they will simply move on. Such abilities have always been essential for the survival and evolution of wildlife in the extremely unpredictable Australian landscape. Experimentation, opportunism and nomadism lead birds to try new or unusual sources of potential foods (so they are ready to take advantage of an unexpected glut of something nice). Towns and cities are full of unusual and sometimes abundant sources of food for those animals willing and able to exploit them. These sources include roadkill, discarded junk food, garbage bins and skips, landfills and school yards as well as the parks and gardens where people *intentionally* provide food for wildlife. And, of course, our backyard bird feeders.

FEEDERS REPRESENT AN UNUSUAL FOOD SOURCE WHEN
COMPARED WITH MOST NATURAL FEEDING SITES
AVAILABLE TO BIRDS. THEY ARE USUALLY PERMANENTLY
LOCATED IN ONE PLACE, AND THE FOOD SUPPLY
IS PREDICTABLE AND RELIABLE. FEEDERS ARE ALSO
OFTEN SMALL IN SCALE, AND COMPETITION FOR THE
AVAILABLE FOOD IS ALMOST CERTAIN.

As we've already seen, feeders represent an unusual food source when compared with almost all the natural foraging sites available to birds. Feeders are usually permanently located in one place, and the food supply is remarkably predictable and reliable. Feeders are also typically small in scale. This combination of features means that competition for the available food is almost certain, although the consequences can vary greatly. Very often, it's simply the birds with the biggest bodies who barge their way to the front of the queue, with everyone else more or less forced to wait their turn. Actual fighting tends to be minimal – but then so does sharing. If the supply is limited, the bigger birds get the greatest amount and whatever is left after they have gone is squabbled over by the next in line.

Two features will influence the nature of the competition and the distribution of the food: the feeder's capacity and the amount of food it makes available. In general, the larger the space, the greater the supply. And the more regular the timing of the food offering, the more intense the level of competition and conflict among visiting birds. Let's explore this a little by considering two very different scenarios.

Firstly, picture a small flat-topped platform feeder where a modest handful of food is placed, once a day, always at the same time each day. This attracts a range of local birds that keep an eye on the feeder and are first to feed, with the usual arrangement of larger species getting in first and smaller species getting what is left. Typically, and perhaps surprisingly, feeders with small amounts of food (in this case, seed) tend to have more birds visiting for shorter periods of time. For some reason the birds take only a small amount of food on each brief visit, and more species get a chance to have a bite. It's almost as though they know that the limited offering is hardly worth fighting over (although this can change if the supply of natural foods is low). This type of feeding regime can have small numbers of birds of numerous species coming and going throughout the day, with relatively little drama.

On the other hand, consider a fairly large expanse of backyard lawn, liberally sprinkled with grain, pieces of bread and bits of meat. The food is replenished several times each day. Each time it is scattered about, a deafening collection of lorikeets, cockatoos, crows, ibis, magpies and pigeons swarm in, fighting and elbowing their way across the now overcrowded lawn to the food. Dozens of other birds sit around the perimeter of the site, on fences and washing lines and in trees, waiting for the chance to join the melee. Most of the birds that are feeding seem to stay for long periods and remain in the vicinity even if they have fed for long periods.

These two vastly contrasting scenarios are both common throughout the suburbs of this country. It is pretty obvious that one will have rather less of an influence on the local bird community, while the other is undoubtedly going to have significant ecological and behavioural effects. It is also unsurprising that a food-induced tsunami of birds is going to have consequences for the human community well beyond the house yard in which it occurs. This type of impact is dealt with in 'Conflict' later in this chapter.

Providing any food for wild birds will change the interaction among species and individuals. A number of studies have demonstrated that the introduction of feeders into an area can dramatically alter the types of species in that location, as well as their abundance. Predictably, these bird communities are often dominated by large, behaviourally assertive species that consume much of the available food, while many of the smaller species usually miss out. In some places, the dominant birds are introduced species such as feral pigeons, common (or Indian) mynas, spotted doves or house sparrows. These species are not what most people hope to attract to their feeders, although some of the larger native birds (such as crows, ibis and cockatoos) are also unwelcome, especially if they come in groups, scaring away the smaller birds.

PROVIDING ANY FOOD FOR WILD BIRDS WILL CHANGE
THE INTERACTION AMONG SPECIES AND INDIVIDUALS.
THE INTRODUCTION OF FEEDERS INTO AN AREA CAN
DRAMATICALLY ALTER THE TYPES OF SPECIES IN THAT
LOCATION, AS WELL AS THEIR ABUNDANCE.

WHAT TO DO

Providing food in a concentrated space at predictable times is always going to lead to some degree of competition, and perhaps quite aggressive interactions, among visiting birds. To minimise this problem, it is better to provide less food rather than more. Increasing the supply only exacerbates the competition by attracting more birds. Remember that most people are attempting to attract birds in order to see them up close; causing excessive aggression is never acceptable. Feed so that you get to see the birds clearly, but not in stressful circumstances.

DOMINANCE IN SUMMARY

🕊 All feeding influences the local bird community and almost always leads to competition for food.

🕊 Bigger birds obtain more food and often prevent smaller species from accessing the feeder.

🕊 Offer small amounts; you are not trying to provide all the food the birds need.

INTERLOPERS

SIGNIFICANCE: 🦅🦅🦅🦅🦅

Everyone has their favourites, but there are also birds we would rather did *not* visit our feeders. These less-wanted species typically include the usual introduced species (feral pigeons, spotted doves and mynas) as well as a number of large and overly dominant natives such as rainbow lorikeets, cockatoos and currawongs. If you get a significant number of any of these birds in your backyard, the small or shy ones don't have much of a chance.

But it is not only the feathered interlopers coming for the bird food that are cause for concern. The presence of numerous birds in a specific location, regularly and predictably, may also attract predators. In reality, they are visiting for the same reason: the opportunity to consume food items present at a feeder. But this time, it is the feeding birds themselves that are the target.

There has been virtually no research on these issues in Australia (as usual), but elsewhere studies have found that both domestic cats and a wide range of birds of prey are attracted to bird feeders. While minimising the impact of the birds of prey can be difficult (they are incredibly fast and can patrol vast areas), the stealth-hunting behaviour of cats can be addressed by giving thoughtful consideration to the positioning of your feeder. The general principle is to place the feeder relatively close to thick, shrubby vegetation; this offers the birds shelter if they need to quickly escape from a predator. However, the feeder should not be positioned too close to dense plantings, as cats may be hiding there hoping to ambush visiting birds. About 2 metres between feeder and shelter is considered a good distance.

Other mammals may be attracted to the feeder, although usually under the cover of darkness. Many feeders in the country are visited by possums (mainly common brushtails), as well as rats and mice. These night-time marauders, which are usually there to clean up leftover seeds or meat, tend to leave abundant droppings behind. This can be an important source of harmful bacteria and also provides a fertile substrate in which disease can grow and spread. The presence of these unseen and uninvited visitors is one more reason why keeping feeders clean is critical (see '*Really* cleaning your feeder' in chapter 3).

These concerns also relate to animals attracted to the debris and leftovers that accumulate on the ground beneath a feeder. And of course, plenty of people deliberately spread bird food directly onto the ground. Such places are ideal attractants for rodents and can support significant populations that live (often undetected) in the garden or in nearby structures.

CONSIDER THE POSITIONING OF YOUR FEEDER CAREFULLY, TO MINIMISE DANGER FROM CATS AND POSSIBLY EVEN CARNIVOROUS BIRDS. PLACE THE FEEDER RELATIVELY CLOSE TO THICK, SHRUBBY VEGETATION THAT OFFERS SHELTER TO BIRDS ESCAPING FROM A PREDATOR – BUT NOT TOO CLOSE, AS CATS MAY BE HIDING THERE HOPING TO AMBUSH VISITING BIRDS. ABOUT 2 METRES FROM COVER IS A GOOD DISTANCE FOR THE FEEDER.

WHAT TO DO

Feeders attract animals that want to eat the food pro-vided – which can include the feeding birds them-selves. Think about safe locations that provide easy access to cover should a predator show up, and always consider that ultimate suburban sit-and-wait ambush predator, the cat. Whether it's a wandering tom, your own 'wouldn't hurt a fly' moggie, or the ninja feral that is rarely seen, never underestimate the intelligence of cats or the impact they have on our birds.

And, again, as always, clean up the scraps regularly to prevent a build-up of vermin. They are out there, whether we see them or not.

INTERLOPERS IN SUMMARY

- Feeders also attract predators and scavengers.
- Uninvited visitors such as vermin and possums can increase the risk of disease and infection.
- Clean the feeder regularly.

CONFLICT

SIGNIFICANCE: 🍁🍁🍂🍂🍂

While there is often very visible conflict among the birds visiting our feeders, the conflict referred to here involves people – people who will usually have very different perspectives on bird feeding. The scenario described earlier in 'Dominance' is by no means exaggerated. Every suburb in every town seems to have places where particular people – presumably trying to be helpful – provide large amounts of food for wild birds.

One of the most familiar scenarios is that of the solitary individual carrying a large bag full of bread to distribute to a vast, squabbling aggregation of ducks, swans and geese near a pond in a city park. I have seen this in every country I have visited. In these cases, any human conflict is usually with the park manager – although they mostly seem to turn a blind eye. It should be fairly clear to anyone who has read this far that this form of feeding is not appropriate (there are many serious issues associated with duck feeding, some of which are discussed in 'Bread and angel wing' in this chapter). But let's consider instead this scale of feeding conducted in a suburban backyard. Conflict ahead!

Some particularly spectacular examples come to mind. One involved an elderly woman who lived immediately adjacent to a coastal national park in southern Queensland. Each weekday, by arrangement, a local bakery delivered all their unsold bread (between 20 and 50 loaves) to her home. At dawn the following morning, all of this was tossed onto the grassed verge next to the park boundary, an activity that regularly attracted around 30 brush-turkeys, who feasted on the bread until they were engorged. Many of these birds remained nearby throughout the day, and their incessant scratching over every inch of ground in the vicinity – including several neighbours' yards – finally removed every molecule of vegetation and leaf litter. Unsurprisingly, the area also supported an extraordinary density of rats. Despite many complaints to the local authorities, this practice continued for over a decade and only ceased with the death of the protagonist.

Another much more traditional example involves the mass feeding of magpies in a large suburban backyard. For over ten years, the elderly male resident has provided an almost continuous supply of beef mince to a vast army of magpies. The numbers fluctuate from around 20 to well in excess of 100. A high proportion of these birds are young unmated magpies, recently expelled from their parents' territories. (These so-called 'tribes' of yearlings – and some older unmated birds – often congregate in such places as they represent safe havens from the harassment of neighbouring territorial males.) The abundant supply of meat has also attracted a range of other species including butcherbirds, crows, kookaburras and even ibis. Needless to say, the issues associated with this many birds remaining for prolonged periods in a relatively small location are extreme, and have inevitably led to conflict. The feeder's neighbours have reported excessive amounts of droppings on house walls, fences, vehicles and even washing, in addition to the offensive smell and the presence of vermin – all of this resulting in reduced property values. Despite the ongoing threat of legal action, the feeding at this site remains active at the time of writing.

BIG BIRDS HAVE BIG APPETITES AND DROPPINGS TO MATCH; THE IMPACT OF LARGE NUMBERS CAN BE VISUALLY (AND OLFACTORILY) SIGNIFICANT. IT IS NO SURPRISE THAT MASS FEEDING AND ITS CONSEQUENCES HAVE LED TO MANY HEATED ARGUMENTS, EVEN TO LEGAL DRAMAS.

The final example also centres around a single species, the sulphur-crested cockatoo. This huge bird – historically found mainly in the drier inland areas – has become increasingly abundant in many Australian cities. Periodic visits to the coast by large numbers of these and other typically rural species – including galahs, cockatiels and crested pigeons – are fairly common during prolonged droughts, but these influxes are usually temporary. Over recent decades, more and more of these birds have found our cities to their liking and have moved in permanently, with 'cockies' among the most conspicuous. Plenty of people who enjoy visits from smaller parrots have had the experience of their modest feeder being overwhelmed by these noisy giants. In one such case, the somewhat bewildered householders were initially encouraging. They provided additional seed, leading to even more cockatoos screeching in each morning. After several weeks of this influx, and an increasing number of complaints from neighbours concerned about the noise, the owners decided enough was enough and simply stopped putting out any seed. The reaction of the cockatoos was to begin systematically chewing any nearby wood, starting with the verandah railings and then moving on to the eaves and fascia of the house, and nearby sheds and fences, before expanding their destruction to include the neighbours' houses as well. Whether this behaviour was the result of the birds simply becoming bored while waiting for their food, or – the preferred interpretation of the people involved – intentional and malicious, the outcome was the same: extensive (and expensive) property damage and ongoing conflict with the people next door.

MASS FEEDING – ATTRACTING LARGE CONCENTRATIONS OF BIRDS TO LOCALISED AREAS WHERE EXCESSIVE AMOUNTS OF OFTEN POOR-QUALITY FOOD ARE CONSTANTLY AVAILABLE – REPRESENTS THE VERY OPPOSITE OF SUSTAINABLE, ETHICAL AND LOW-RISK BIRD FEEDING.

Each of these scenarios is, believe it or not, remarkably common all over this country. These are peculiarly Australian problems, related mainly to the size of the birds involved. If the species flocking in were the size of the common feeder birds from the northern hemisphere, such as blue tits and chickadees, no one would even notice. I just can't see a major suburban conflict over a large flock of even really nasty zebra finches or silvereyes.

BREAD AND ANGEL WING

Bread is wonderful stuff, for humans. It is simply not suitable for any other species, even though just about everything will eat it. Bread in itself is not particularly harmful; it's just bulky, and doesn't provide much in the way of nutrition – especially the blindingly white tasteless version. This is still (inexplicably) the most common sort of bread purchased, and therefore the most widely used to feed birds, both at home and in parks, with many negative results. If birds eat excessive quantities of bread, they fill up without gaining much benefit. In addition, massive build-ups of bread in ponds cause ecological problems, encouraging algal growth and leading to a decline in water quality.

Just as worryingly, a very serious condition known as 'angel wing' is associated with birds that consume lots of bread, something that occurs among the ducks and other waterbirds at ponds all over the world. These birds are often unable or unwilling to procure natural foods, and their excessive intake of bread (as well as chips, biscuits and cake) results in a carbohydrate-rich diet with insufficient levels of vitamins D and E, calcium and manganese. Angel wing appears to be related, at least in part, to a bread-dominated diet. Excessive amounts of carbohydrates and protein, with insufficient calcium and vitamin E intake when young, causes the 'wrist' section of the wing to develop abnormally. This leads to the deformed wings that are characteristic of the condition. In affected birds, certain long wing feathers grow twisted at a strange angle, often sticking out perpendicular to the body. This prevents flight and ensures that the birds are unable to leave the area – thereby making them even more dependent on consuming unsuitable foods provided by people. Angel wing is incurable and often leads to early death in the affected animals. This is just one more reason to avoid using bread as bird food.

Mass feeding and its consequences can lead to the involvement of councils, community groups and police, and very real psychological trauma can result. Currently, there appear to be few legal avenues that either side of such conflicts can pursue, with most cases coming down to some form of voluntary mediation. They are very rarely resolved to anyone's satisfaction. This contentious aspect of the complex landscape of wild bird feeding is one of the most potent reasons for opposition to feeding in this country.

What about the perceptions and beliefs of the people who are providing food for these large groups of birds? This issue is almost completely unstudied, but my own interpretation, based on conversations with many people at the centre of these disputes, is that they genuinely believe they are helping the birds, and see their actions as a form of welfare. 'They wouldn't be coming if they weren't hungry' is a common assessment. The ensuing conflict with neighbours and authorities is therefore unfortunate, but perhaps inevitable if one 'has to do the right thing'. In other words, this type of feeding is a *moral* activity, with any opposition being by definition *amoral*. These perspectives are difficult to counter.

The reality is, however, almost completely different. Feeding to attract mass concentrations of birds, in localised areas where excessive amounts of often poor-quality food are constantly available, represents the very opposite of the sustainable, ethical and low-risk bird feeding I am hoping to encourage. Virtually every legitimate concern discussed in this chapter about major bird feeding issues is amplified by this form of feeding. Dependency, disease, poor nutrition, aggression, predation and conflict with people are all exacerbated. Morally, and in many other ways, such a practice is wrong.

WHAT TO DO

Resolving these conflicts will not be easy. If you are someone who has been feeding what might be regarded as 'too many birds', reading this book may give you a different perspective. It is not the feeding per se, it is the amount of food. Of course, you don't want to shut this down instantly; that would certainly be unethical. Do so gradually, over a few weeks. Some of the birds may protest and get stroppy, but sooner or later they need to fend for themselves.

If you are someone (such as a neighbour) affected by birds aggregating because of excessive feeding, you probably already assume that talking is fairly point-less. Maybe it is time to try again. You may strongly disagree with what's going on, but the people feeding the birds often sincerely believe that what they are doing is right – it is important to respect that and try to see things from their perspective. But they also need to appreciate that their actions are probably exposing their much-loved birds to harm. Hopefully, the information in this book, sensitively shared, will help in this process.

CONFLICT IN SUMMARY

🐦 Mass feeding, whether in public or at home, is always a bad idea.

🐦 While attracting large numbers of birds with continuous food supplies almost certainly increases the likelihood of bad outcomes, it is the resulting conflict between feeders and other people – especially neighbours and authorities – that is of particular concern here.

🐦 Personal intervention, with genuine respect, is the best approach.

3

FEEDERS

SETTING THE TABLE

How you set the table depends on who's coming to dinner. Or at least who you hope turns up. There's no use sticking up a hanging tube feeder like the ones you saw in London or New York or Amsterdam unless there are small seed-eating species with sharp beaks in your area. And there's not much point in erecting those Austrian-style 'little house' feeders on a pole you snapped up at that garage sale if you're trying to attract king parrots or kookaburras. The feeder (and the food) you use must match both your location and the species you are encouraging to visit.

Although plenty of people simply toss out whatever they are using for bird food onto the back lawn, most offer their fare on or in some sort of feeder, a physical apparatus that provides a place to display the food and where birds can stand or perch. Feeders come in many designs, but they can be classified into several basic types. Only some of these are used in this country – with platform feeders being by far the most common. However, the increasing interest in wild bird feeding in Australia, and the growing availability of feeders designed in other countries (for very different birds) in shops and online, makes it worthwhile discussing most of these feeder styles.

LOCAL CONDITIONS CAN CHANGE, AND THERE MAY BE SENSIBLE REASONS TO PUT OUT LARGER AMOUNTS OF FOOD. BUT THESE CIRCUMSTANCES ARE ALMOST ALWAYS TEMPORARY. WHEN THE CONDITIONS RETURN TO NORMAL, SO SHOULD THE QUANTITY OF FOOD YOU OFFER.

The main categories of feeders are based on the manner in which the food is provided or dispensed, or the type of food being presented. Many different names are used for these feeders, but I will apply the following common labels:

- platform feeders
- tube feeders
- seed feeders
- hopper feeders
- nectar feeders
- suet feeders.

As we shall see, there are variations and special cases that don't fit neatly into the above categories, but these tend to be fairly unusual. Each type has a particular purpose and has features that need to be considered when we are thinking about feeding. Before we describe the feeders themselves, there are three crucial questions that apply to all feeders:

- How much food can (or should) they provide?
- Where should they be positioned?
- (and, most importantly) How easy are they to clean?

HOW MUCH FOOD?

This is one of the fundamental issues (apart from whether we should be feeding at all) associated with feeding wild birds, and it applies to every type of feeder. Effectively, all feeders offer an endless and – if replenished regularly – guaranteed supply of human-provided food. This may seem ideal if we are worried about the availability of natural food resources, especially during harsh or prolonged adverse weather conditions (drought, bushfires, long cold periods, the aftermath of severe storms). Obviously, local conditions can change, and there may be sensible reasons to put out larger amounts of food. But these circumstances are almost always temporary. When the conditions return to normal, so should the quantity of food you offer.

Under normal circumstances – based on what we now know about the foraging ecology of many well-studied species – we should *not* be trying to provide all, or even a sizeable proportion, of the birds' diet. In most situations, our feeding is and should be just a supplement to their natural diet. And don't be guided by the fact that the birds are eating everything you provide. Except in certain unusual situations, it doesn't mean they are hungry.

This applies particularly to large feeders. Just because you have the space to supply plenty of food (and it will almost certainly all be consumed) doesn't mean that you should. This very feature – an apparently never-ending supply of food – may actually be one of the main problems with feeders, of every type. This becomes an important personal issue for you to consider: just why are you feeding in the first place?

Here are a few suggestions for regulating the amount of food you provide. As always, it is ultimately up to you and your reading of the local conditions.

- Provide only a small amount of food each day.
- If you really don't know what might be appropriate, try supplying just a handful.
- Never provide more food than could be consumed in a morning by all visiting birds.
- Be aware that excess food can become spoiled or mouldy if it gets moist.
- Don't continually replenish the feeder – allow it to stand empty at times.

WHERE SHOULD THE FEEDER BE PLACED?

There are two main factors to consider when deciding where to install a feeder: visibility and vulnerability. There may be a complex array of motivations for feeding, but most people want to be able to see the birds as clearly as possible when they come to visit. Obviously, this means that feeders need to be placed in locations where they can be observed easily, often from inside the house. Unobscured viewing also allows you to scan for potential threats (cats, snakes) and to notice accidents, such as a bird becoming caught or somehow entangled (sometimes a toe can become stuck in the structure). These events are pretty rare, but they can happen and you might need to respond quickly.

FEEDING THE BIRDS AT YOUR TABLE

POSITION FEEDERS IN LOCATIONS WHERE YOU CAN
OBSERVE THEM EASILY, USUALLY FROM INSIDE YOUR
HOUSE. UNOBSCURED VIEWING ALSO ALLOWS YOU TO
SCAN FOR POTENTIAL THREATS (SUCH AS CATS, OR
SNAKES) AND TO NOTICE AND RESPOND TO OCCASIONAL
ACCIDENTS, SUCH AS A BIRD BECOMING TRAPPED.

The visibility of the feeder should also be considered from the perspective of the visiting birds. Placing food on a conspicuous structure out in the open can expose birds to the risk of predation, especially from cats and birds of prey. Cats are more likely to attack from a nearby place of seclusion, such as a dense thicket, while hawks typically launch rapid surprise raids in relatively open spaces. In both cases, smaller birds will try to escape into any nearby foliage cover. It is therefore good to have some dense planting close to the feeder, but not close enough to increase the risk of feline attack. Most of the advice I have read suggests that locating the feeder about 2 metres from cover is optimal.

Some organisations also point out the risks associated with window-strike, a much bigger issue than most people realise, affecting birds all around the world. The risk of window-strike will depend on lots of features unique to your particular house and yard (or apartment and balcony), so it is difficult to make general suggestions. If you suspect that nearby windows represent a hazard for birds visiting your feeder, definitely change things.

With the exception of those that are permanently attached to a structure, feeders should be moved around periodically. This minimises the possible build-up of discarded food refuse, and droppings, on the ground beneath. Selecting a new location will obviously require consideration of all the factors just discussed, each time.

CAN THE FEEDER BE CLEANED EASILY?

This is a vital question, but the answer varies enormously. Each feeder type offers different challenges that are discussed in the box '*Really* cleaning your feeder'. The principle is nonetheless the same for all feeders: they must be clean at all times, and that involves the daily processes of brushing, washing and drying, as well as periodic but regular bleach cleaning. Even if the feeder looks uncontaminated, clean it anyway.

Now, let's look at the various feeder types.

REALLY CLEANING YOUR FEEDER

Of all the activities involved in feeding wild birds, this is probably the most important. Keeping the feeder as clean as possible is the key to preventing and controlling disease. No matter how or what you feed, you need to develop a simple but regular cleaning routine so you can be confident that wherever you are offering food, the surface or container is always clear, clean and dry.

If you are using a platform feeder – any flat, horizontal surface – prior to placing the food, you need to sweep, wash, wipe and dry it, every time! For daily cleaning, you can wash with cleaning vinegar, if the surface is not too dirty. If it's pretty bad, move on to bleach. And, if the feeder design allows it, you need to dismantle and cleanse the whole thing thoroughly on a regular basis. The most effective and complete (and proven) treatment involves the following steps:

- Scrub all feeder parts with a firm brush and hot soapy water (dishwashing detergent is fine), ensuring that all waste material (old food, droppings) is removed.
- Bleach, by immersing the feeder components in a mixture of 1 part liquid chlorine to 9 parts water for a few minutes. Domestic bleach – typically 5 per cent sodium hypochlorite – is suitable. Never use bleach undiluted.
- Dry completely, ideally in the sun. Wiping with paper towel can speed up the process.

Note that you should not use the scrubbing brushes and other cleaning items (rubber gloves, cloths) for any other purpose. Store these items outside the house. Always wash and dry your hands thoroughly before and after cleaning the feeder.

PLATFORM FEEDERS

The 'platform' refers to the horizontal surface upon which the food is placed. This can be square, octagonal, oval, circular or just about any other shape. However, a rectangular platform is by far the most common. If it has legs or a central stand, a feeder like this is typically called a 'bird table'. Platforms can come in virtually any dimensions, from dinner plate to surfboard size. Although commercial versions are now available in shops and online, in Australia most are still homemade or repurposed. I have seen circular metal rubbish-bin lids, large pot-plant trays, birdbaths (without the water), aluminium baking trays and – yes – actual surfboards, all used as platform feeders.

Platform feeders are suitable for any food type. Their defining feature is that the food is simply placed on top. Easy. The birds, just as simply, land right on this surface and start eating. But obviously this means that the birds will be walking about on the eating surface and may defecate there as well. You can see the potential problems this poses, I hope. And the larger platform feeders provide sufficient space for lots of food and lots of birds. That may be just what you want, but it does mean an even greater chance of aggression between the visitors and increased risk of infection should there be sick birds among them.

LARGE PLATFORM FEEDERS PROVIDE SUFFICIENT SPACE
FOR LOTS OF FOOD AND LOTS OF BIRDS. THAT MAY BE
JUST WHAT YOU WANT, BUT IT DOES MEAN AN EVEN
GREATER CHANCE OF AGGRESSION AND INFECTION.

Platform feeders are typically either hung, or fixed to a pole or structure. Hanging feeders can be suspended from ropes or wires, usually attached to each corner, which are secured to a tree branch or a purpose-provided metal stand. This style of feeder is usually far from stable, moving about in the wind and when birds land and thus feeders are more likely to allow food to be lost over the side unless there is a raised edge. These edges, however, make them a lot more difficult to clean thoroughly. In general, suspended platform feeders are more suited to smaller species such as finches, and the littler parrots and lorikeets.

Fixed platform feeders are immobile, and include the traditional bird table on legs, or platforms that have been attached to the top of a pole, verandah or balcony railing, or that have simply been placed on a solid structure such as an old 44-gallon metal drum or brick barbecue-style 'box'. This type of feeder can be much larger in size. The biggest I have seen was a reused cattle-feed trough, about 5 metres in length, fixed to a fence, which was situated near the homestead of a rural property: it was used mainly by a vast number of galahs and sulphur-crested cockatoos (being supplied with sunflower and sorghum grown on the farm). The most elaborate feeder I have come across was a homemade metal contraption comprised of seven large circular platforms arranged at various heights and angles along a vertical pole. It resembled a gigantic high-tea cake stand and must have cost a lot to stock with food. But the mass breakfast visitation by native pigeons, doves and rosellas was spectacular.

As mentioned already, the biggest concern with all platform feeders is the almost inevitable mixing of food and faeces. No other feeder design is affected by this problem as acutely; in other feeder types, the birds tend to perch (and crap) away from the food they are eating. As a result, daily thorough cleaning of platform feeders is extremely important. This is explained in detail elsewhere in this chapter (see '*Really* cleaning your feeder'), but must involve removing any leftovers and droppings, then washing the surface (with vinegar or soapy water) and carefully drying it before putting out any new food. I can't emphasise this enough: if you have a platform feeder, it must be cleaned before use, every day. If you can't guarantee this process, don't use a platform feeder or don't feed at all.

IF YOU HAVE A PLATFORM FEEDER, IT MUST BE CLEANED BEFORE USE, EVERY DAY. IF YOU CAN'T GUARANTEE THIS PROCESS, DON'T USE A PLATFORM FEEDER OR DON'T FEED AT ALL.

This crucial issue raises the matter of the material used to construct the platform. Generally, the easiest materials to clean are smooth and sheer: metal, fibreglass or hard plastic, which can be easily swept and washed, and which dry quickly. Unfortunately, most platform feeders are made of wood (especially the DIY versions), which is the worst material to use. Unless it has been painted for the purpose (with non-toxic all-weather paint), wood is porous, stays damp and is extremely difficult to keep clean. The fibres and minute cracks and indentations that are almost always present on wooden surfaces are ideal places for nasty bacteria and viruses to reside. These features also make wood almost impossible to clean completely and very slow to dry. Best to avoid wood altogether. These issues actually relate to spaces, corners, edges and other structural features in *any* feeder where debris could build up. The key aim is to ensure there are no places where malevolent nasties can hide and flourish.

Some platform feeders have a roof over the feeding surface. This is mainly an attempt at keeping the rain off but is often not very successful. The roof structure can also deter large species from visiting.

TUBE FEEDERS

These are by far the most common type of feeder used throughout the world, but are still relatively unusual in Australia. This is because they were designed mainly for the many small granivorous species in the northern hemisphere that feast on seeds: tits, chickadees, robins and a wide variety of finches. The only equivalent species we have here are our native finches, as well as several introduced birds such as the goldfinch, greenfinch and house sparrow, which are still common in some places. Most of the smaller species we may be lucky enough to have living nearby tend *not* to be granivores. Fairy-wrens, silvereyes, honeyeaters, flycatchers, wagtails and gerygones, for example, are all primarily insect-consumers – some of these birds also seek out pollen and nectar when it is available seasonally.

BIRDS USING A TUBE FEEDER HAVE TO BE FAIRLY SMALL IN SIZE AS THEY NEED TO PERCH ON A VERY SHORT ROD OR RING.

Tube feeders are designed to supply a lot of seed over a long time in small amounts. The tube is almost always in the form of a transparent perspex, hard plastic or UV-stabilised polycarbonate cylinder (the latter is becoming increasingly common), allowing both birds and people to see how much and what sort of food is available. These feeders can be filled with almost any of the usual seed types and mixes, except for 'straights' (a single seed type rather than a combination) of very small and slippery seeds such as nyger; these require a specialised feeder design.

DISEASE TRANSMISSION IS OF MUCH LESS CONCERN WHEN USING TUBE FEEDERS BECAUSE THERE IS A REDUCED CHANCE OF DROPPINGS CONTAMINATING THE FOOD. HOWEVER, SOME SERIOUS OUTBREAKS OF AVIAN DISEASES IN THE NORTHERN HEMISPHERE HAVE BEEN TRACED TO DROPLETS EXPELLED THROUGH THE BEAKS AND NOSTRILS OF INFECTED INDIVIDUALS EATING AT THESE FEEDERS. IT IS CRUCIAL THAT YOU THOROUGHLY CLEAN ALL TYPES OF FEEDERS.

Birds using a tube feeder have to be fairly small in size as they need to perch on a very short rod or ring protruding from the tube, directly below a circular hole that provides a way to get at the seeds within. There are usually pairs of these perches and matching holes on opposite sides of the tube. Depending on the length of the tube, there may be up to four pairs of perches and holes; in theory, this would limit the number of birds perching and feeding at any one time to a maximum of eight individuals. In practice, this is about right – for a particular micro-second. In reality, the displacement of feeding birds by others wanting a turn is almost continuous. Studies – using high-speed video – of the activities of flocks of chickadees attending tube feeders have shown that individuals only get a few seconds to feed every time they visit, so each must return continuously. Of course, this is mainly a problem when large numbers of birds are attempting to feed together.

Disease transmission is of considerably less concern when using tube feeders because there is a much lower chance of droppings contaminating the food being consumed. The arrangement of the perch and the fact the seed is accessed through a hole only a little larger than the bird's head seems to minimise this contact. Nonetheless, some very serious outbreaks of avian diseases in both North America and the United Kingdom have been traced to droplets expelled through the beaks and nostrils of infected individuals. It is still crucial that you clean tube feeders, though perhaps not as often as is required for platform feeders. Most of the recently manufactured tube feeders are designed explicitly with cleaning in mind: they are easily deconstructed so that every component can be safely and easily washed and dried. Older models are not so conveniently designed, and it requires considerably more effort to dismantle them and ensure that every little space and surface is thoroughly cleansed. Whenever possible, all feeders need to be dismantled and bleach-cleansed regularly.

SEED FEEDERS

Seed feeders offer birds access to particular varieties of bird food that cannot be dispensed by either the standard tube feeders or the larger hopper style described below. They allow birds to feed on seed types that are either too large or too small for the usual gravity-fed feeders. Seed feeders closely resemble tube feeders, consisting of a hanging, see-through cylinder that contains the food. This cylinder is constructed of either metal mesh or hard plastic mesh, with perforations of an appropriate size to allow birds to peck at the seeds within. No perches or holes are needed, as the birds simply cling on to the mesh and insert their beaks through the spaces.

The main food types used in these feeders are whole peanuts and nyger. Peanuts are among the most important of all bird foods throughout the world and are used in a large number of seed mixes. Whole peanuts are the largest of all bird foods and are widely regarded as being potentially dangerous for many birds, especially if offered to nestlings. Apocryphal stories of chicks choking on a whole peanut circulate everywhere in the feeding world, although reliable direct observations of this phenomenon are extremely rare. Nonetheless, because of such fears peanuts are often manipulated into much smaller pieces – 'granules' or 'splits' – before being added to seed mixes. Whole peanuts are still enormously popular (with birds and the people who feed them), so the acceptable safe method of providing them is in an appropriately sized mesh feeder. Birds land on the mesh and 'nibble' on the peanut within. This might seem finicky, but birds everywhere love these nuts.

Nyger (a seed from an African daisy) represents the opposite problem, as it is too small to be offered in most feeder designs. The sleek, ultra-smooth surface and flattened elliptical shape of these tiny, shiny seeds enables them to slide effortlessly through the feeding holes of a typical feeder and fall (wastefully) onto the ground below. Again, the solution has been to contain

WHOLE PEANUTS ARE THE LARGEST OF ALL BIRD
FOODS AND ARE REGARDED AS BEING POTENTIALLY
DANGEROUS FOR MANY BIRDS, ESPECIALLY IF
OFFERED TO NESTLINGS.

the seeds in a metal mesh cylinder – in this instance, mesh with extremely fine apertures. This means that only a very small number of species are able to access these tiny seeds through the mesh. In the United Kingdom, goldfinch and greenfinch are among the few birds that can use this type of seed and feeder, and as a result they have virtually no competition. In the case of the greenfinch, the dramatic growth of their populations from much lower levels in the 1980s has been attributed directly to the arrival of nyger in the garden feeders of Britain. Tragically, the recent catastrophic decline in the species was found to be largely associated with a disease that spread from the very feeders they were so happily exploiting.

Feeders offering singles of either peanuts or nyger are rarely if ever used in Australia at present. If it was found that some of our smaller grain-loving bird species liked these foods, this situation could change significantly.

HOPPER FEEDERS

This refers to a wide variety of feeder types that feature a substantial storage component, known as the hopper; the hopper can hold a large amount of seed, which it makes available more or less continuously. Typically, a feeding tray is attached to the base of the storage unit, which is continually replenished with food as it is consumed. Because the hopper can contain a significant amount of food and be topped up as often as needed, this type of feeder is effectively perpetual – it never seems to run out. For this reason, hoppers are widely used to feed cage birds and chickens.

Hopper feeders are usually quite heavy and so are more likely to be placed on a rigid structure, but they can also be suspended. They dispense a very wide array of seed types and mixes but are unsuited for dispensing straights of very small seeds such as nyger, canary, millet and canola; these would simply flow out of the opening at the bottom of the hopper. (Unless included in mixes with larger grains, these seeds need to be offered in specially designed tube feeders.)

Hopper feeders tend to have a relatively small feeding zone, typically allowing only a limited number of birds to perch and feed at the same time. This can encourage greater turnover of visitors, and shorter feeding times. This design also minimises the chances of food and faeces mixing; the arrangement of the perch and feeding tray usually means that a bird's rear end is facing away, and droppings normally fall to the ground below. Obviously, this is valuable for maintaining cleanliness and reducing the risk of transferring infection.

Like any of the feeders, hoppers do need to be cleaned regularly. On the other hand, hopper feeders are often of a fairly robust design and may require quite a bit of effort to clean. Ideally, the whole thing needs to be taken apart, but for most of the designs I have seen this is difficult or impossible. The concern this raises is not so much related to disease transfer as the possibility of spoiled grain becoming stuck inside the apparatus and contaminating any new food placed within.

NECTAR FEEDERS

This category of feeders covers a wide array of devices that provide a way for birds to obtain various forms of liquid food. They can be as simple as a saucer of sugar-water or homemade dispensers made from plastic bottles, all the way through to elaborate imported hummingbird feeders that resemble flowers. (But unless you imported the hummingbirds as well, these are not going to have many visitors!)

FEEDING HOMEMADE SUGAR-WATER TO WILD BIRDS IS BECOMING LESS POPULAR, AS PEOPLE ARE NOW AWARE THAT THE RICH LIQUID CAN RAPIDLY BECOME CONTAMINATED WITH BACTERIA AND YEAST. THERE ARE FAR BETTER COMMERCIAL PRODUCTS AVAILABLE FROM SPECIALIST PET FOOD SHOPS.

Nectar feeders are intended to be used by the sweet-tooths among the local birds – honeyeaters, silvereyes and of course lorikeets, whose natural diet includes the carbohydrate-rich nectar produced by many of our native plants. These species have an uncanny ability to detect the presence of fruit-derived sugars (fructoses), and they will readily find and consume homemade concoctions of sugar or honey dissolved in water. Feeding sugar-water to wild birds has been practised for a very long time but is now far less popular, in part because people are more aware that the rich liquid can rapidly become contaminated with bacteria and yeast. A number of nasty health issues can beset birds that consume the contaminated fluids. This is especially the case in hot weather, when harmful organisms can reproduce rapidly within hours of the liquid being provided. For this reason, nectar feeding should only be practised where you can be absolutely certain that the liquid will be consumed quickly or that the dispenser can be retrieved after a relatively short period (hours not days). Nectar feeders must never be left out for days on end, and must always be very thoroughly cleaned after each use.

Although plenty of people still make their own sugar or honey mixtures, there are far better commercial products available from specialist pet food shops. These are usually in the form of powders, which are dissolved in water prior to being poured into the feeder (sugar-water in relation to feeding honeyeaters and nectarivores is also discussed in chapter 4).

SUET FEEDERS

Suet refers to a wide range of solid products made almost entirely of animal fat, although the array of ingredients now used to supplement the fat is quite extensive. Traditionally, the suet used for feeding birds was simply the lard derived from the household kitchen. Today, dozens of versions are produced in the northern hemisphere, primarily for feeding birds during the long harsh winters. Obviously, this is not such an issue in Australia (although I recently heard of blackbirds and black-faced cuckoo-shrikes using imported suet blocks in Tasmania). I include them here for completeness' sake.

WHAT BIRDS NEED TO EAT

Whether their diet is normally grasshoppers, berries, fish or chips, all animals (including us) require the same basic components: proteins, fats, carbohydrates, vitamins and minerals. A balanced diet must have all of these, though not necessarily every day. Birds will often binge on food that is seasonally abundant (fruit, flowers, caterpillars) or eat opportunistically, for example when a dead kangaroo, insect infestation or overturned wheat truck turns up. But eventually their body will let them know when something is missing. We know now that birds are astonishingly clever at detecting minute amounts of what they need, even in types of food they don't necessarily recognise.

There are also times of the year when these nutritional demands are especially powerful. The build-up to breeding is one of the most important. Females, in particular, must prepare their bodies for making eggs, a task that requires enormous amounts of time and energy and large quantities of particular molecules. Protein is vital, as well as calcium and various minerals for constructing the shells. For males, additional protein may be needed to build up muscles for the prolonged singing and fighting that occurs at the start of the breeding season. At such times, birds will seek out foods rich in these components, and even species that specialise in fruit, nectar or seeds may consume insects or meat.

The feeders are simply the containers that hold the solid block of suet. These are usually in the form of small metal or plastic cages with fairly large openings, or sometimes mesh bags. They are attached to tree trunks or hung from branches; the birds simply land and eat the stuff through the gaps.

As wild bird feeding develops and matures in Australia, we are likely to see an increased use of different feeder designs, reflecting a desire to attract a much wider range of species to our gardens. This will also be accompanied by an expansion of the types of food being offered.

4

FOOD

WHAT'S ON THE MENU?

Planning your menu depends on who you expect to be coming. It is pretty simple really: if you want magpies, don't serve vegetables; if you hope for finches, you will need to find out what they like. And if you want to make it interesting and would like to attract a wide variety of species, you will need a variety of offerings (served in a variety of places). If we are going to do this seriously, sensibly and sustainably, it is time to go beyond meat and vegies (or, in this case, mince and birdseed).

You would be forgiven for thinking that feeding birds is just a matter of plonking something edible on the bird table or in the feeder. That is always going to be your main daily activity, but we also need to appreciate that this is actually a major ecological experiment. Naturally, we are all focused on our own backyard, garden or feeder, supplying attractive morsels in an attempt to bring in those free-flying visitors. We tend not to see our private spot at a larger scale. Yet if we took a drone-view of our street, suburb or town, and were able to see all the other feeders in all the other backyards, we would realise that there really are a lot of places that the birds can visit. We might be oblivious to all these other foraging opportunities, but the birds certainly aren't. Clever studies using tiny transmitters attached to the birds and special detectors set up at feeders have revealed astonishing levels of traffic between feeders, with some individual birds visiting a dozen over the course of a morning. Rather than your garden being the one place where 'your' birds come to get all their sustenance, your feeder is one small part of a huge and complex network of invisible bird pathways spreading over sometimes vast areas.

I think it is important to get this picture in our heads as we think about the food we are providing. We now know that this simple practice can have far-reaching consequences in terms of the health and wellbeing of the birds we welcome. We also need to know that what happens at our place is somehow connected to many other feeders too, whether we like it or not. It's just one more reason we all need to be much more thoughtful about the way we practise our feeding.

There are lots of ways that this section on foods could be presented. I have decided the most useful approach is to focus on some general categories of birds, based on their natural diet (you've already seen a summary in chapter 1, 'Australian feeder birds grouped by diet'). These categories are not entirely exclusive but they should cover the majority of species that are likely to visit feeders in Australia. (If you have experience with successfully feeding species not included here, let me know.)

STUDIES USING TINY TRANSMITTERS ATTACHED TO BIRDS AND SPECIAL DETECTORS SET UP AT FEEDERS HAVE REVEALED ASTONISHING LEVELS OF TRAFFIC BETWEEN FEEDERS, WITH SOME BIRDS VISITING A DOZEN TIMES OVER THE COURSE OF A MORNING. YOUR FEEDER IS ONE SMALL PART OF A HUGE AND COMPLEX NETWORK OF INVISIBLE BIRD PATHWAYS SPREADING OVER SOMETIMES VAST AREAS.

The best approach – in terms of optimal nutrition and doing the least harm – is to try to mimic the sorts of natural foods the birds would be seeking themselves. Like us, birds will happily eat things that are radically different from what they should be consuming. Like us, they could be adversely affected if they ate too much of certain things (really, how much sugar or fat or salt *should* you consume in a day?).

Another general principle already mentioned is that it is a good idea not to feed at the same time every day. This will be hard to do, simply because we are all creatures of habit. But it is a natural extension of the idea of mimicking what happens in nature: very rarely is the same food available in the same place every day. So, as much as you can, be intermittent in your feeding, as well as springing a few surprise foods.

In this section I will make a lot of suggestions and recommendations, but these are almost all just informed guesses at the moment (although I did talk to a lot of experts). Until Australian researchers have investigated the many unknowns associated with the feeding of wild birds in this country, this is the best we can do.

FEEDING PARROTS AND COCKATOOS, PIGEONS AND DOVES

SPECIES COMMONLY USING FEEDERS IN AUSTRALIA:

GALAH, SULPHUR-CRESTED COCKATOO, KING PARROT, EASTERN ROSELLA, PALE-HEADED ROSELLA, SPOTTED DOVE, CRESTED PIGEON

This is a large and untidy group of moderately large birds whose diet is predominantly or entirely made up of seeds, mostly obtained while feeding from the ground – although some also forage a lot within the foliage of trees. The collective list of the plants from which they obtain their diverse daily foods is enormous, as it includes large numbers of native and introduced grasses, shrubs and tree species. Some of the parrots that visit feeders also have a fondness for fruit, blossoms, nectar, certain leaves and twigs and quite often even insects and their larvae. Generally, the natural seeds they consume tend to be small, and it can require quite a bit of manipulation to get to the edible stuff inside the seeds' 'packaging'. Many of these birds also take advantage of the paddocks full of wheat, oats, barley, canola, sorghum and sunflowers that are grown throughout the country. Almost all of these commercial seeds are much larger than those obtained naturally; they provide more nutrients per unit and are usually much easier to 'open'. Virtually all of these commercial grains feature in the wild bird seed mixes available in Australia.

Before going on to actually suggest what we might want to offer parrots and cockatoos, it is important to note that wildlife veterinarians and nutritionists completely avoid a seed-only diet for captive or reha-bilitating birds. Instead, they recommend a diet con-sisting mainly of special pellets (see below) with plenty of greens and fruit. Although this relates to birds completely dependent on humans for their food, the point is that the usual mixes available in the shops do not provide an adequate diet for Australia parrots. We need to ensure that we are only offering the birds visit-ing our feeders a fraction of their daily diet. We want them to continue to find most of their food elsewhere.

WHAT TO OFFER

There are a variety of seed mixes available in pet shops and supermarkets. Although many have 'wild bird' marked prominently on the packaging, in most cases these mixes are identical to those developed for cage birds. It is hardly surprising that the mixes used for parrots, cockatoos, pigeons and doves contain a

THE MOST POPULAR BIRDSEED IN THE KNOWN UNIVERSE: SUNFLOWER

Ah, yes, sunflower seeds! Just about every bird everywhere prefers sunflowers to anything else. This is primarily due to the fat content; in fact, the oil contained in this plant's seeds is the main reason that sunflowers are grown in such enormous quantities around the world – primarily intended to be sold for human consumption. The oil content of the black sunflower is among the highest of any plant seed, typically about 40 per cent. Birdseed sunflower has a much lower oil content – and its seeds (which are usually striped) are therefore less attractive to birds than the high-oil black varieties. As a result, almost all birdseed mixes include some black sunflowers, even though too much of this sort of oil is definitely not a good idea. Indeed, because lots of birds so clearly adore these super-oily seeds, plenty of people provide black sunflowers alone on their feeders. And plenty of companies, knowing that they can sell this stuff literally by the truckload, are only too willing to sell you 'singles' of black sunflower. This is a serious issue – too much oil is very bad for all birds. Some of the more responsible manufacturers of wild bird mixes in Australia have recognised the health risks of high-oil sunflowers and are actually starting to reduce the proportion of these seeds in their mixes. This should be our approach as well.

lot of seed types, given the enormous breadth of their natural diet and the fact that many of these birds already consume commercial grains out on the farms. However, if we follow the principle of trying to provide food that resembles their natural diet as closely as possible, the optimal mix should consist of plenty of smaller sized seeds – such as the various millets, canary seed (from a type of grass) and nyger – along with the usual sunflower seeds.

Australia's parrots have evolved with a diet that is relatively low in fat, a reflection of their natural food sources. This has led to an associated nutritional problem. When parrots eat lots of high-fat oilseeds such as sunflower, canola and safflower, the resulting 'flush' of protein and fat may cause a range of significant health problems (although it can be useful in preparing for breeding). This is another reason to limit the amount of these seeds in the food that we provide to wild birds.

THE SWARMS OF GALAHS AND COCKATOOS DESCENDING ON WHEAT AND SORGHUM CROPS AROUND AUSTRALIA TYPICALLY ONLY EAT THESE SEED TYPES WHEN THEY ARE STILL SOFT AND 'MILKY', BEFORE THEY HAVE HARDENED AND ARE READY FOR HARVEST.

As a general recommendation, it is best to select commercial seed mixes that are produced by pet or animal food companies (see 'Buying wild bird food in Australia' in chapter 1) rather than the generic – though cheaper – packages produced for the supermarkets. Many of the latter are composed of high proportions of cereals such as wheat, oats and sorghum, which are not that attractive to most of the native birds listed here (in contrast to the feral pigeon, which appears capable of eating almost anything). This may seem strange given the swarms of galahs and cockatoos descending on wheat and sorghum crops around the country. However, the birds typically only eat these seed types when they are still soft and 'milky', a week or two before they have hardened and are ready for harvest. The small rock-hard billiard balls of sorghum found in many seed mixes are very different from young sorghum seeds with their doughy consistency.

A serious problem associated with most dry seeds, which have been heat-treated so that they can't grow, is that the little nutrition they do provide in the way of vitamins and minerals is very difficult for the birds to absorb; in other words, the 'bioavailability' is low. This is not the case with so called 'live' seed – which can be hard to find. For this reason, most wildlife

ITEMS FOR FEEDING PARROTS AND PIGEONS

PROVIDE PLENTY OF THESE

- ✓ Cockatiel/Small parrot mix
- ✓ Parrot pellets
- ✓ High-quality, high-variety seed mix
- ✓ Millet, canary seed, nyger
- ✓ Wheat
- ✓ Peanut pieces
- ✓ Grapes
- ✓ Blueberries
- ✓ Pieces of fruit (apples, pears, melons, chilli, passionfruit)
- ✓ Native plant seeds
- ✓ Native plant blossoms

PROVIDE IN MODERATION

- ± Safflower seeds
- ± Sunflower hearts and pieces
- ± Oats (note: susceptible to fungal growth if it gets wet)

AVOID THESE

- ✗ Bread
- ✗ Black sunflower seeds
- ✗ Striped sunflower seeds
- ✗ Desiccated coconut
- ✗ Anything salty
- ✗ Anything sugary (including honey)
- ✗ Anything fatty
- ✗ Whole corn
- ✗ Sorghum (also called milo)
- ✗ Never citrus

vets recommend specially prepared 'parrot pellets' for captive parrots, and these would also be good for wild species.

The best commercial seed mixes that can be used for this group are becoming increasingly diverse in their contents. They can include:

- various dried fruits and berries
- pieces of peanuts and other nuts
- native seeds such as wattle and eucalyptus
- specifically designed pellets containing certain important nutritional elements.

You can expand on this theme by adding things like grapes, blueberries, bits of apple and melon, and even blossoms from plants in your (or the neighbour's) garden. But never include any citrus; far too acidic.

FEEDING LORIKEETS

SPECIES COMMONLY USING FEEDERS IN AUSTRALIA:
RAINBOW LORIKEET, SCALY-BREASTED LORIKEET,
MUSK LORIKEET

Although lorikeets belong to the same family as the parrots mentioned earlier (the Psittacidae), they are categorised separately here because of their defining physical adaptation, the brush-tongue, which enables them to deftly extract pollen and nectar from the blossoms of a very broad range of Australian plants. This specialisation has allowed them to prosper spectacularly in our towns and cities because of the ubiquitous plantings of nectar-bearing species such as callistemons, grevilleas and banksias. Nonetheless, their diet is much broader than pollen and nectar; they consume blossoms, fruit, berries and a wide variety of seeds. They also eat a surprising number of insects, especially caterpillars. Unexpectedly, lorikeets are also fond of the occasional bite of raw meat (often munching on roadkill; see 'Extreme diet switching: seeds and meat' in chapter 6) and it has recently been discovered that many lorikeets partake of the mince being offered to magpies everywhere in Australian suburbs. Unlike the previous group of birds, however, lorikeets forage primarily within the foliage of trees – they rarely feed on the ground – often in noisy, conspicuous, squabbling groups.

THE LORIKEET'S DEFINING PHYSICAL ADAPTATION IS ITS BRUSH-TONGUE, WHICH ALLOWS IT TO EXTRACT POLLEN AND NECTAR FROM THE BLOSSOMS OF A VERY BROAD RANGE OF AUSTRALIAN PLANTS. STILL, THEY DO EAT PLENTY OF SEEDS.

WHAT TO OFFER

Although lorikeets spend most of their time foraging among flowers for pollen and nectar, they do eat plenty of seed (contrary to a quite pervasive urban myth) as part of their very broad natural diet. This does *not* cause them to lose those remarkable brushes on their tongues as is often claimed. But they can't live on these sugar- and carbohydrate-rich substances alone. Finding commercial products to provide for lorikeets is a snack compared with sourcing food for most other groups of wild birds in Australia, as there are plenty of high-quality choices. My suggestion is to provide foods as diverse and interesting as these outrageous extraverts themselves.

In terms of food types suitable for lorikeets, we can start with the same varied seed mixes described for the previous group. Lorikeets tend to be much less likely to partake of the larger, harder seeds like sorghum and corn but, like everyone else, they do love sunflower seeds. Again, too much of these high-oil items is not a good idea. But when it comes to what a lorikeet will happily munch on, the variety is extraordinary (as shown in the section 'Items for feeding lorikeets').

Lorikeets are extremely popular among backyard aviary keepers, and a great deal of research has been conducted into the diets of captive lorikeets. This has resulted in a plethora of lorikeet-specific food products now being available for their wild relatives. Some of these products include small pieces of fruit and vegetables (such as apples, carrots, spinach and figs), sultanas, currants, various berries and all sorts of nuts (usually in bits). As well as these packages of loose seed (and other things) mixes, there are blocks of all of the above ingredients bound together by edible gums, starches and honey. These blocks or tubes of more-or-less solid material can be placed directly on the platform feeder or bird table or hung from branches. This form of solid bird food is becoming increasingly popular and lorikeets really do seem to like them. (I can also confirm that a wide variety of

ITEMS FOR FEEDING LORIKEETS

PROVIDE PLENTY OF THESE

- ✓ Lorikeet food (wet or dry)
- ✓ Various 'nectar' products
- ✓ High-quality wild parrot seed mix
- ✓ Pieces of fruit (apples, pears, melon, mango)
- ✓ Green and leafy (lettuce, spinach, broccoli)
- ✓ Pieces of vegetables (carrots, sweet potato, pumpkin, green beans)
- ✓ Grapes
- ✓ Strawberries
- ✓ Native plant seeds
- ✓ Native plant blossoms

PROVIDE IN MODERATION

- ± Black sunflower seeds
- ± Unsweetened fruit juices (apricot, peach)

AVOID THESE

- ✗ Bread
- ✗ Acidic fruits (oranges, lemons, mandarins)
- ✗ Avocado
- ✗ Chocolate

species that normally avoid my feeder – blue-faced honeyeaters, pied currawongs, spangled drongos and others – do visit almost every time I stick one of these blocks on the platform.)

A general warning concerning any of the sugary foods is that they can easily become contaminated with bacterial and yeast outbreaks, especially in summer. Make sure food containers are not left out for long periods and always clean them thoroughly.

FEEDING LARGE INSECTIVORES/CARNIVORES

SPECIES COMMONLY USING FEEDERS IN AUSTRALIA:

AUSTRALIAN MAGPIE, PIED BUTCHERBIRD,
GREY BUTCHERBIRD, LAUGHING KOOKABURRA

I realise that is a pretty clumsy way to categorise these big 'meat'-eaters but it will do the job for now. A few other species might be included as well: pied currawongs and your local corvid species (crows and ravens). All of these birds naturally consume lots of invertebrates, especially worms, grubs, spiders and beetles, as well as small skinks and snakes. Some people argue that the term 'omnivore' might be a better label. Although I find that just too general to be meaningful, it is more than fair to say that their diet extends well beyond this modest list and that, given the opportunity, they are likely to try most potentially edible things. All of these species are known to scoff down barbecued sausages (sometimes directly from the fork en route to the mouth), unguarded patty cakes at children's birthday parties and pet food intended for a household dog or cat. Clearly, they are not very fussy when it comes to food choices.

WHAT TO OFFER

When it comes to this particular group – which includes the most popular species that people feed – it is extremely important to get things right. Although all of these birds will consume any of the meaty human foods – mince, ham, salami, devon, fried chicken, heart – as well as things like cheese, bread, chips, cake and an almost endless array of junk food, this most definitely does not mean that they should.

If we follow the principle of attempting to provide food that is as close as possible to their natural diet, we need to look for high-protein items with plenty of calcium. That instantly excludes most things listed above; served alone, any sort of meat is just too low in calcium. Crucially, this means that mince is out altogether, unless – and I mean *only if* – you add digestible calcium power (half a teaspoon for every cup of mince) or one of the commercial invertebrate supplements. These supplements are now widely available from most of the big pet stores and some veterinarians. You can also add things like rice cereal, breadcrumbs and special insectivore mix (although this will not change the calcium levels). The idea is to substitute for the nutritional elements that are missing from raw, processed meat.

Another approach is to prepare a homemade 'dense porridge' that can be spooned out onto a platform feeder as required. One recipe (recommended by a zoo keeper) combines finely chopped raw chicken frames (yep, bones, cartilage, the lot), raw eggs, Weet-bix™, insectivore mix and a little grated cheese. This will obviously take some preparation, but if we are serious about nutrition and minimising the risks associated with feeding these birds, it is a great idea.

As we have already seen, the reason that mince is the most common food item offered to magpies is simply its convenience. Realistically, any mince replacement needs to be readily available and not too expensive. The most likely candidate would therefore seem to be commercial pet food, which is formulated for dogs and cats – mammals rather than birds –

but is usually well balanced nutritionally (although, again, some vets worry that the iron levels may be a bit high for birds). Pet food products can be found in any supermarket and corner store in the country, so availability is not an issue.

While admittedly not yet exhaustively researched, it is possible that even chopped-up bits of pet meat roll, or 'dog sausage' – with the supplement added – are more suitable for magpies than a pile of sticky mince. Another option could be the chunkier types of tinned dog food (although any wet foods will make it harder to clean the feeding surface). My numerous veterinarian correspondents were generally reluctant to endorse pet food as genuinely suited for magpies, but when pushed suggested that dry puppy kibble would be their preferred option. (Some people add just a little water, but the vets say dry is best. The opinion of the magpies may be another matter.) It's definitely worth a try.

Of course, you could avoid the artificial, processed stuff altogether and offer actual live and wriggling worms. You could dig these up from the garden or gather them from your own worm farm. This is being practised by some serious magpie fanatics around the country, although they say it's hard to keep up the supply. Certain pet food companies are currently attempting to manufacture artificial worms for the potentially massive magpie market, but until these appear on the supermarket shelves we need other sources. Perhaps the most natural of the alternatives are mealworms, either live and wriggling or dried and not. Quite a few people have made the switch to these beetle larvae already. Mealworms are available from certain specialist pet stores (and also online), but they are unlikely to become the most popular mince substitute. In addition, some experts are concerned that mealworms are too fatty to be recommended as a major part of the diet.

WHEN FEEDING INSECTIVORES AND CARNIVORES, CONSIDER AVOIDING ARTIFICIAL, PROCESSED FOOD AND OFFERING LIVE AND WRIGGLING WORMS. THIS IS BEING PRACTISED BY SOME SERIOUS MAGPIE FANATICS AUSTRALIA-WIDE, ALTHOUGH THEY SAY IT'S HARD TO KEEP UP THE SUPPLY. YOU COULD DIG UP THE WORMS FROM THE GARDEN OR GATHER THEM FROM YOUR OWN WORM FARM.

If you take my message seriously and decide to change from mince to, say, pet food, remember that

ITEMS FOR FEEDING LARGE INSECTIVORES/ CARNIVORES

PROVIDE PLENTY OF THESE

✓   Pet food (dry or soaked – vets say dry – or chunky wet types)

✓   Mealworms

✓   Pet meat roll (dog sausage) mixed with insectivore food

✓   Commercial insectivore food

✓   Mince *only if* mixed with calcium powder equivalents

PROVIDE IN MODERATION

±   Grated cheese

±   Meat roll (dog sausage)

AVOID THESE

✗   Any raw meat including mince

✗   Bread

✗   Processed meats (salami, German sausage)

✗   Cooked bones

✗   Anything fatty

✗   Anything sugary

you will need to be patient. It can take a while for the birds to adjust to the alternative. After all, the new stuff will look nothing like mince. Start by mixing a little of the substitute into the food you put out and then gradually increase the ratio. It's definitely worth the effort for our beloved maggies.

FEEDING HONEYEATERS AND NECTARIVORES

SPECIES COMMONLY USING FEEDERS IN AUSTRALIA:

BLUE-FACED HONEYEATER, RED WATTLEBIRD, NEW HOLLAND HONEYEATER, LEWIN'S HONEYEATER, EASTERN SPINEBILL, SILVEREYE, NOISY MINER (INEVITABLE)

This category includes many more species than the few listed here; these are just the more common ones recorded nationally. Every location will have its own array of honeyeaters – the largest family of birds in Australia and among the most typical 'bush birds' in the country. Despite their group name, these birds almost certainly consume more insects than nectar, although all do sip their fair share of the sweet stuff. Like lorikeets, silvereyes are equipped with a specialised bristly tongue designed to mop up nectar and pollen from flowers. Nonetheless, all of these birds also eat a lot of insects as well as slurping up the juice from grapes and ripe fruit.

DESPITE THEIR NAME, HONEYEATERS PROBABLY CONSUME MORE INSECTS THAN NECTAR AND POLLEN – THEY ARE ALSO PARTIAL TO FRUIT, ESPECIALLY IF OVERRIPE.

WHAT TO OFFER

This group of birds is most easily catered for by offering foods as similar as possible to those in their natural diet. In fact, you can provide *exactly* what they consume in the wild: flowering native plants that supply nectar and pollen. When it comes to this particular feeding group, the best foraging opportunities can be provided simply by planting the right sorts of shrubs and trees. And that is what has been happening on a massive scale throughout Australia for decades; spectacular cultivars of callistemons, grevilleas and banksias are being grown everywhere. As we now know, this has not resulted in more species visiting our gardens. All that nectar attracts nectarivores all right: large numbers of noisy miners, and in some places, wattlebirds and blue-faced honeyeaters – and, of course, rainbow lorikeets (see 'Managing miners' in chapter 5). These are wonderful native species, but they are usually intolerant of any other birds that may compete for these resources. Or, in the case of noisy miners, any other species up to the size of a wedge-tailed eagle!

You are almost certainly well aware of these issues by now, having learned the hard way about the unintended consequences of your planting choices. There is a significant rethink underway to determine how

**ITEMS FOR FEEDING HONEYEATERS
AND NECTARIVORES**

PROVIDE PLENTY OF THESE

✓   High-quality commercial nectar mix

✓   High-quality commercial insectivore food

✓   Grapes and stone fruit

✓   Overripe fruit (nectarines, peaches, bananas)

PROVIDE IN MODERATION

±   (Note: I originally included sugar-water here but the vets
     were adamant; only very small amounts are acceptable)

AVOID THESE

✗   Bread

✗   Jam, marmalade

✗   Baked goods

best to design gardens so they attract a greater diversity of species, especially the smaller species that have been less able to cope with the bird bullies. There is still plenty we can do to create perpetual foraging opportunities, and the importance of integrating habitat (rather than just 'living food') is explored in some detail in the next chapter.

To attract this group of birds to a feeder, we also need to think beyond the sugar trap. All of these species have traditionally been offered various forms of sugar-water, usually simple homemade concoctions of jam, honey, molasses, sugar or marmalade dissolved in water. This is poured into dishes, saucers or sometimes DIY sugar-water dispensers. The birds definitely flock to these offerings, and vast amounts of the sticky stuff is consumed. As already discussed in the section on nectar feeders in chapter 3, the practice is now strongly discouraged because of problems associated with the consumption of too much sugar and contamination by bacteria and yeast. It is possible to provide these homemade products safely, but only if the quantity is small and the feeding vessel is always cleaned thoroughly.

These significant concerns can be avoided altogether if you provide one of the numerous high-quality commercial nectar mixes, especially if supplied in combination with similarly high-quality insectivore foods. A quick online search will reveal a remarkable variety of these products.

Finally, all of these species are partial to juicy fruit such as grapes and stone fruit, especially if overripe (at least by human standards). Don't throw away those oozing black bananas or leaking nectarines – silvereyes and spinebills will love them!

FEEDING FRUCTIVORES (FRUIT-EATERS)

SPECIES COMMONLY USING FEEDERS IN AUSTRALIA:

EASTERN KOEL, FIGBIRD, SATIN BOWERBIRD, SOME
RAINFOREST PIGEONS

A much smaller group, although in some places in Australia species such as fruit-eating pigeons and bowerbirds will be attracted to feeders with fruit offerings.

WHAT TO OFFER

Feeding these birds is largely a DIY affair: offer whatever you can buy or gather locally, whole or in pieces. This could include overly ripe grapes, blueberries, figs or stone fruit. You could also gather up fallen native figs and palm tree fruit if they were plentiful.

ITEMS FOR FEEDING FRUCTIVORES

PROVIDE PLENTY OF THESE	AVOID THESE
✓ Small native fruits	✗ Bread
✓ Grapes, blueberries	✗ Jam, marmalade
✓ Pieces of stone fruit, apple, melon	✗ Citrus (yes, even though they are fruit)

FEEDING SMALL INSECTIVORES

SPECIES COMMONLY USING FEEDERS IN AUSTRALIA:

GREY SONG-THRUSH, WILLY WAGTAIL, FAIRY-WRENS,
FANTAILS, FLYCATCHERS, WHISTLERS

These are definitely not the types of birds you would
expect to see at backyard feeders, but it is possible.
Most of these species have not done well in the face
of urbanisation, so any considered attempt to assist
them to survive among us is worthwhile.

WHAT TO OFFER

The key to attracting these insect-eating species into
your garden is a naturally occurring supply of insects.
Although it appears that the majority of the food
items these little birds consume are taken in the air,
they also glean and sally in and around the right sorts
of foliage. This takes us into the realm of garden habitat, which is discussed in the next chapter, but the
general idea is to plant native shrubs and small trees
that support a wide array of insects and thereby provide food for the birds to consume. It's a rather indirect form of feeding birds, but also gets us thinking
beyond the feeder; we need to consider providing a

ITEMS FOR FEEDING SMALL INSECTIVORES

PROVIDE PLENTY OF THESE	AVOID THESE
✓ Overripe fruit	✗ Bread
✓ Commercial insectivore food	✗ Jam, marmalade

more complex space for these small and often sensitive birds to live in.

There are commercial products available for insectivorous birds and these can certainly be offered. All of them have been developed specifically for raising insect-eating species in captivity, for example in zoos, or for use during the rehabilitation of injured birds. These products are therefore of the highest quality and are fairly easy to obtain from specialist vets or pet stores, as well as online. It is also a good idea to offer these food items at a completely different location from any other feeders in your garden.

Some people have been able to support certain insectivorous birds by putting out insect-attracting items such as rotten bananas, figs and custard apples. I have seen grey fantails, willy wagtails and superb fairy-wrens mopping up the fruit flies and tiny wasps hovering over a mound of overripe native figs piled onto a feeding table.

FEEDING FINCHES

SPECIES COMMONLY USING FEEDERS IN AUSTRALIA:

DOUBLE-BARRED FINCH, ZEBRA FINCH, RED-BROWED FINCH, DIAMOND FIRETAIL, CHESTNUT-BREASTED MANNIKIN

These tiny, highly social gems are among the groups of birds most significantly impacted by urbanisation and associated threats like cat predation. Habitat clearing and alteration, weed and grass control (both spraying and mowing), and fire management activities have fragmented and degraded the places typically occupied by these birds; these interventions have also greatly disturbed their breeding areas. Above all, the birds' traditional foraging resources – expanses of native and introduced grasses and weeds – have been dramatically affected. What local councils see as 'tidying up', and as essential for controlling fire risk and snake populations, has been catastrophic for the many species of finch that used to live among us. Although they do respond well to feeders offering the small seeds they love, the bigger challenge is habitat.

WHAT LOCAL COUNCILS SEE AS 'TIDYING UP', AND AS
ESSENTIAL FOR CONTROLLING FIRE RISK AND SNAKE
POPULATIONS, HAS BEEN CATASTROPHIC FOR THE MANY
SPECIES OF FINCH THAT USED TO LIVE AMONG US.

WHAT TO OFFER

As with the small insectivores, providing some form of
dense foliage is important for these vulnerable birds
– although for the finches, this is primarily for cover
rather than food. Having said that, there are a number
of native grasses that can be planted specifically as
a food source for these birds. If you have the space
in your yard, these are really the ideal way to feed
finches; you are supplying exactly what they would be
feeding on naturally. It would be sensible to ask a local
nursery or catchment group about what grass species
are most suited to your particular location.

In terms of the seeds to offer on a feeder, a wide
variety of commercial mixes is available. As usual,
these will have been developed with captive granivorous birds in mind, most obviously Budgerigars,
canaries and lots of native finch species. These mixes
typically comprise various millets and pannicums,

which all originate outside Australia but are entirely suitable for our finches. Other commercial seeds include nyger and phalaris. Thankfully, a remarkable array of native grass seeds are being produced in commercial amounts, although for growing rather than for feeding to birds. Still, this opens up a wonderful new dimension in offering native seeds to native finches.

Because there are significant numbers of native finches in captivity, there is plenty of advice online about the sorts of things these birds like to eat. Live food in the form of mealworms is regularly mentioned, as are greens, specifically lettuce. Worth a try.

THERE ARE A NUMBER OF NATIVE GRASSES (ESPECIALLY 'WALLABY' AND 'KANGAROO' GRASSES) THAT CAN BE PLANTED SPECIFICALLY AS FOOD SOURCES FOR FINCHES – YOU WILL BE SUPPLYING EXACTLY WHAT THEY WOULD BE FEEDING ON NATURALLY. ASK A LOCAL NURSERY OR CATCHMENT GROUP WHAT GRASS SPECIES ARE MOST SUITED TO YOUR LOCATION.

ITEMS FOR FEEDING FINCHES

PROVIDE PLENTY OF THESE

- ✓ Planted native grasses
- ✓ Commercial finch seed mix
- ✓ Millet (white French, Japanese and shirohie)
- ✓ Pannicum (yellow and red)
- ✓ Nyger
- ✓ Greens (lettuce)
- ✓ Mealworms

PROVIDE IN MODERATION

- ± Hulled oats

AVOID THESE

- ✗ Bread
- ✗ Sorghum

WHAT CAN WE FEED THE DUCKS?

Despite its universal popularity, feeding bread to the ducks and other waterbirds down at the local pond is actually a bad idea. The problem is that bread – and donuts, chips and cake – is not suitable food for birds and can lead to all sorts of physiological and ecological problems (discussed in 'Bread and angel wing' in chapter 2).

Is it possible, then, to enjoy this interaction with wildlife and not cause harm? Thankfully, the answer is 'Yes'. Increasing numbers of parks and other locations where waterbirds are fed by the public are suggesting that alternatives to bread can be used. Certain places sell little bags of these healthier food options, and some authorities have set up exchanges where you can swap your bread for alternative items.

Provided there are no DON'T FEED THE DUCKS signs nearby, you can offer any of the following with a clear conscience. These are all harmless, provide some decent nutrition and may be beneficial to the birds. But remember, never offer too much, especially if you don't know what quantity the birds have already consumed.

Bread substitutes:
- Lettuce, cabbage, broccoli and spinach (torn into small pieces)
- Corn (tinned, frozen or fresh)
- Rolled oats (uncooked)
- Rice (brown is best; cooked not raw)
- Peas (defrosted or raw)
- Duck pellets (manufactured specifically for domestic waterfowl)

5
HABITAT

IT'S MUCH MORE THAN FOOD

Birds need a place to live, not just something to eat. In reality, most of the time they don't need our food, but they cannot do without an appropriate physical space in which to carry on with their lives. Obviously the relatively small area represented by our backyard, courtyard or balcony is not going to provide everything a bird might require to survive and reproduce. But it can offer a safe and reliable place to stop and visit during their normal daily routines. If it included some useful snacks, a place to drink and bathe, and somewhere to rest out of sight, it is much more likely that visiting birds will be back again tomorrow. It may even be a suitable location to build a nest.

The key idea of this brief chapter is that we need to start thinking well beyond the feeder and the bird food and begin to consider all the other critically important things that birds need, especially in the suburban landscapes where most of us live. For many of the species that visit us, their main 'home' will always be any existing natural bushland nearby or further away. These birds spend plenty of time in the strange and artificial world of the human because there are numerous valuable resources to exploit. Because people like variety and colour, suburbs almost always contain a much larger assortment of plants than you'll find in natural landscapes. This is particularly pronounced when it comes to flowering shrubs and small trees: our gardens and parklands are bulging with callistemons, banksias and grevilleas as well as lots of spectacular flowering eucalypts, often originating far away in other parts of Australia. Plenty of these plants are cultivars with bright and conspicuous blooms developed to be over-the-top gaudy. Colourful plants from overseas are widely used as street trees and in urban parks: African tulip tree, jacaranda and poinciana come to mind.

The abundant supply of nectar and pollen produced by a lot of these plants has resulted in a massive increase in the numbers of birds such as rainbow lorikeets, noisy miners, wattlebirds and others. The irony is that these plants were (and continue to be) marketed as 'bird-friendly' and 'honeyeater magnets'. 'Bring in the birds! Fill your garden with colour!' trumpet the labels, failing to mention that the most likely outcome will be the arrival of clouds of big, noisy species, while smaller birds seem to disappear altogether. The reason we have got to this point is, again, because of food. When we (and this included me) decided to grow these plant species in our gardens, we were actually introducing just another type of human-provided food, this time in the form of enhanced super-blossoms at ridiculously high densities – not that different from setting up a thousand plates of sugar-water. It is time to completely rethink how we design Australian gardens if we want to benefit more than just a few big nectar-obsessives.

THE POPULARITY OF PLANTS MARKETED AS 'BIRD-FRIENDLY' (INCLUDING CALLISTEMONS, BANKSIAS, GREVILLEAS AND FLOWERING EUCALYPTS) HAS RESULTED IN A MASSIVE INCREASE IN BIG, NOISY SPECIES SUCH AS RAINBOW LORIKEETS, NOISY MINERS AND WATTLEBIRDS.

MANAGING MINERS

There have been big changes in the abundance and species of birds found in Australian suburbs over the last few decades. A lot of the smaller species such as finches, fairy-wrens, flycatchers and small honeyeaters have declined, primarily because of the destruction of their habitat. At the same time, some species have boomed, with rainbow lorikeets and noisy miners among the most obvious. While any increase in numbers might be welcome, when it comes to noisy miners we have a real problem.

From one point of view, these native honeyeaters are just highly successful at living in the suburban environment. The reason for this is the huge numbers of grevilleas, callistemons and other nectar-bearing shrubs and trees that we have planted in our gardens and parks. Miners love all that sweet stuff. But they refuse to share it, driving out virtually all other species, including many that don't even like nectar. This has meant that our selection of garden plants – often chosen with the goal of attracting more species – has actually resulted in more miners but less everything else.

We can't remove the miners but we *can* provide cover for the smaller species, in the form of dense shrubs and understorey. We can also provide food appropriate to these species, such as small seeds for finches and mealworms for the insect-eaters.

GARDENS NEED TO BE HABITAT

Some people think that our preference for living in a savanna-like setting goes way back to humankind's origins in the open landscapes of East Africa, regions that featured a few scattered trees with wide swathes of grassland in between. There are places in Australia that look just like that – for example the broad tropical savannas of Queensland and the Northern Territory. But the modern urban version has more trees and is a lot greener: the typical Australian park or backyard is a flat expanse of green lawn with sometimes a tree or two. The only low vegetation tends to be small sparse shrubs in linear garden beds growing along the fenceline. This leaves a wide area of grass with some shrubs at the edges; ideal for a game of backyard cricket but fairly useless as wildlife habitat. Wander through the conservation park nearby (because there are reserves in and around almost all our towns and cities) where the natural bush remains and what you will notice – in addition to the plants looking quite 'untidy' – is the different layers of foliage at all levels: clumps of grasses and small plants growing along or close to the ground (amid the leaf litter and fallen branches); dense hedges of shrubs and young trees; and layers of leafy vegetation at a variety of heights.

It will be uneven in height and structure and maybe chaotic, but far more varied and complex than most gardens. Obviously, the details will vary enormously depending on where you live and the climatic zone, but the complexity is usually pretty evident.

> NATURAL HABITAT IS AN ECOSYSTEM OF INFINITE
> PARTS AND INTERACTIONS, PROVIDING RESOURCES,
> OPPORTUNITIES AND SUBSTRATES FOR ALL OF THE
> SPECIES PRESENT.

In these habitats you will find birds that are able to forage for food at each of the levels, from the ground (and just below the surface) to the very tops of the trees. Some tend to spend most of their time at particular heights (in very simplistic terms, certain finches, robins and fairy-wrens at the bottom; flycatchers and silvereyes in the middle; thornbills and honeyeaters at the top) and others seem to be happy foraging at any level. Watch closely and you can see how busy and meticulous they are, searching behind every leaf and twig, probing every flower, minutely examining each crack and fissure in the tree bark as they seek insects, fruit, seeds and maybe nesting material. And if danger is detected or something startles them, they are gone in a flash, disappearing into the densest foliage they can find. That is what natural habitat is like: an ecosystem of infinite parts and interactions, providing resources, opportunities and substrates for all of the species present.

Okay, let's not pretend that we can construct something that detailed and complicated by sticking a few plants in the ground. But we can certainly do a lot better than we have done to date. The good news is that enormous progress is being made in the challenging process of restoring and reconstituting habitat – often including areas that have suffered terrible damage and neglect. All over Australia, ordinary but dedicated people are banding together (as catchment and landcare groups, 'friends of' organisations and, at a much larger scale, as members of conservancies – privately owned conservation reserves) in order to bring back some version of a particular location's original ecosystem. These activities often involve huge amounts of volunteer effort and dedication, and there is currently a lot of discussion about what their aims should be: returning the site to a state that resembles what things were like at a particular time in the past; or attempting to redesign something appropriate to the current situation. We don't need to get bogged down in these debates here. What is important for us is that all this work and experimentation has resulted in a lot of valuable information and advice suited to local situations. The 'local' is crucial, because it has become clear that whatever we do at the level of our own backyard must be relevant to the ecology of where we live. For that reason, one of the best things you can do is join your local catchment group. This will not only give you access to an enormous amount of valuable local knowledge, it will reinforce the importance of seeing that your house yard is part of a wider interconnected landscape.

BIRDS NEED COVER

So, what does all this mean for your place? At its simplest, it means thinking about adding structural complexity. This can be in the form of some denser shrubs, the lower level of foliage that is typically missing from most gardens. There are many of these plants available, but the key feature should be the *density* of the living structure. Ideally the shrub should have many little branches with lots of leafy growth close to its inner stems or trunk. The idea is to create a thicket, a place small birds can vanish into when needed.

LANTANA: INTRACTABLE WEED OR VITAL COVER?

Controversially, areas infested with thickets of the invasive shrub lantana actually provide some of the best cover for the smaller bird species, as well as for small mammals – and many butterflies love lantana flowers. This has led to a radical rethink about this noxious weed among the restoration community. The latest approach is to very gradually replace lantana thickets with native shrubs, rather than suddenly removing the lot.

No, I'm not suggesting that you should plant lantana, but hopefully you see the point. If you do have some of this wild and unruly shrub, subdue it slowly and plant natives such as hakeas and some smaller wattles with a view to replacement. Lantana is also susceptible to shading, and in places such as the edges of rainforest it acts as a very effective nursery for native tree species. Saplings of these trees prosper within the gloom of the lantana thicket, nurtured by the thick leaf litter, until they eventually emerge and sprint up into the light. As they grow and their foliage spreads, the lantana below slowly thins and diminishes, although it never completely dies out. This is a process that requires time – but sometimes we need to take the long view.

This might be in order to avoid the neighbour's cat (yours is safely inside, of course) or possibly a hawk, but much more likely is an attack by noisy miners. The unfortunate presence of a colony of these bullies has led to the disappearance of smaller native species (such as fairy-wrens, finches and flycatchers) in many areas. Studies have shown, however, that some of these birds can coexist with noisy miners provided there is a decent cover of dense understorey available to them (see 'Managing miners' in this chapter).

Plant cover that is useful for smaller birds need not only be in the form of low-level shrubs. The zone between lower level foliage and the tree canopy above is also important. If you have the room, include trees of smaller stature – or short-lived specimens that occupy that space.

INSECTS ARE FOOD TOO

So far our discussion of bird foods has included the stuff offered on or in feeders, as well as plant-based nectar and pollen, but plants provide another crucially important type of food: insects. Actually, to be accurate, we should call these organisms 'invertebrates', because plenty of bird favourites such as snails, slugs, spiders, scorpions, centipedes and worms are not actually insects (you know, six legs, etc.). While some birds eat their meat in the form of lizards, frogs and other animals, by far the most important source of protein for the majority of bird species is represented by the invertebrates – this is the case even for birds regarded as nectar specialists, such as honeyeaters, silvereyes and lorikeets. Most of this protein is consumed as caterpillars, larvae, grubs and pupae found on leaves and flowers. (It is also worth saying that even species that exclusively eat grains or fruit as adults raise their nestlings entirely on insects.)

INVERTEBRATES ARE THE MOST IMPORTANT SOURCE OF PROTEIN FOR THE MAJORITY OF BIRD SPECIES, INCLUDING BIRDS REGARDED AS NECTAR SPECIALISTS. MOST OF THIS PROTEIN IS CONSUMED AS CATERPILLARS, LARVAE, GRUBS AND PUPAE FOUND ON LEAVES AND FLOWERS. EVEN SPECIES THAT EXCLUSIVELY EAT GRAINS OR FRUIT AS ADULTS RAISE THEIR NESTLINGS ENTIRELY ON INSECTS.

It may seem strange to some people, but I really am suggesting planting vegetation that attracts insects; yes, insects that will eat the leaves – but which will in turn be eaten by insectivorous bird species. This fairly radical strategy is only at the very earliest stage of development, but a growing list of native shrubs and trees known to support insects is being suggested as suitable for Australian gardens. These include:

- acacias
- baeckeas
- correas
- hakeas
- leptospermum
- melaleucas
- prostanthera.

If we can get this right, it will be a massive advance in our overall aim of bringing many more types of birds into our backyards and lives, using a very different type of bird feeder.

WATER

This is really about as obvious as it gets. Australia is a very dry place most of the time, and access to water is always important for birds. There really is no reason why every house (and apartment, townhouse and mansion) shouldn't have a birdbath or two. That seems pretty straightforward, but even with such an apparently uncontroversial topic as birdbaths, there are key factors that need to be considered: position and cleaning. The same rules apply to both birdbaths and feeders, so I won't go into too much detail again here. However, I will reiterate that the location of the birdbath should be the best compromise between providing a clear view for the humans and foliage cover for the birds so they can find refuge in case of danger.

CONCERNS RELATED TO BIRDBATHS INCLUDE TOXIC ALGAE, AND CONTAMINATED FOODS BEING LEFT TO DECOMPOSE IN THE WATER. YOU SHOULD REGULARLY EMPTY, CLEANSE, DRY AND REFILL YOUR BIRDBATH.

When it comes to cleaning, all the familiar concerns associated with feeders – the concentration of large numbers of birds and different species, faecal contamination and the possibility of spreading infection – are relevant to birdbaths. Theoretically, cleaning should be a lot simpler for a birdbath. Empty, cleanse, dry and refill. Yes, 'dry'. Although birdbaths present fewer dangers associated with disease, there are some significant concerns related to toxic algae, as well as contaminated foods being left to decompose in the water. (Crows and ravens are notorious for leaving bits of food waste in birdbaths.) The best way to deal with most of these problems is to thoroughly clean the container and then allow it to air-dry in strong sunlight.

What makes this otherwise simple task a lot more difficult is the surface structure of the birdbath itself. Many traditional designs are constructed of cement or brick-like material, which are almost always rough and grainy to touch, providing plenty of surface area where algae and bacteria can shelter and breed. These surfaces are extremely difficult to clean properly. Far better are newer materials that are smooth and sheer – and therefore much easier to keep clean. It is important to note that this feature can actually be treacherous for the birds coming to drink and bathe: the slippery surface can make it quite difficult for them to get out, especially when they are a little waterlogged after a bath. If you have this type of birdbath, provide something like a stone or stick that will allow birds to escape easily.

6
FEED

CAN BIRD FEEDING BE ETHICAL,
RESPONSIBLE AND SUSTAINABLE?

I am sticking my neck out by writing this little book. For a large number of people – many of whom I count as friends and colleagues – the idea that anyone could actually be promoting the feeding of wild birds is fairly worrying, maybe even unforgivable. Surely, they say, I must be aware of the many problems and impacts associated with this frankly misguided practice. Hadn't I already concluded in my previous book that providing additional food for wild birds was almost always pointless? Were not the arresting final words of *The Birds at My Table*: 'We think our feeders are for the birds. Our feeders are actually for us'?

All true, kind of. I am certainly acutely aware of the many issues and problems, and I agree that in almost all cases the birds just don't need any extra food.

I want to state clearly that I am not intentionally promoting the feeding of birds. Because I don't need to. Millions of people are already deeply engaged in this activity and don't have to be encouraged. They will continue to feed wild birds, whatever any so-called 'experts' or authorities say. And while I do feel that feeding is usually primarily undertaken for the benefit of the people rather than the birds, I strongly believe that this practice is *not* simply pointless or misguided.

MANY FEEDERS WORRY ABOUT WHETHER THE FOOD THEY PROVIDE IS SUITABLE BUT HAVE NO WAY OF FINDING OUT. THEY WOULD LOVE TO KNOW HOW TO FEED THEIR BIRDS BETTER, WHICH FOOD TO USE AND WHAT THEY SHOULD BE DOING TO MINIMISE GENUINE ISSUES SUCH AS DEPENDENCY AND DISEASE.

No, the reason for my spending a lot of time I don't really have describing an activity that is almost certainly going to land me in even greater trouble, is

EXTREME DIET SWITCHING: SEEDS AND MEAT

One of the strangest and most unexpected outcomes of wild bird feeding in Australia involves magpies and rainbow lorikeets. This is a very Aussie situation: it just couldn't happen anywhere else. Both of these species are now among the most abundant species in urban areas throughout the country and both are fed in a huge number of gardens, backyards and balconies – and sometimes in the *same* location. Obviously, the food being provided for each species is completely different: meat (typically mince) for one and seed for the other. And obviously neither species will be interested in the other's meal. At least, you wouldn't have thought so.

Over the past decade, as rainbow lorikeets have become increasingly common, many have discovered the feeders containing meat put out for the magpies. Now, alarmingly large numbers of rainbows, throughout their range, indulge in this seemingly unsuitable diet. After all, these birds are specialised for pollen and nectar eating, although they certainly eat plenty of seed as well. What is going on?

As we investigated this in more detail, we discovered that – to our great surprise – consuming raw meat is actually a normal, though irregular, activity for lots of parrots. We found records and photographs of parrots munching on road-killed kangaroos, farm-slaughtered animals and barbecued steaks. A lot of people sent us photos of smiling families sitting around the dining table with the disturbing addition of the pet cockatiel enjoying the roast chicken. Disturbing at first, but it turns out that for species that don't have much protein in their usual diet, getting stuck into raw meat when available is essential for preparing for the breeding season. Meat-eating in parrots is apparently 'natural'!

What is not 'natural' is the amount of meat, often in the form of mince, piled onto endless suburban feeding platforms. What was an opportunistic, once-in-a-while event has now become something our lorikeets can indulge in daily. We have no idea what this might mean in the long term but I have a nasty feeling that it is not likely to be good.

that I believe strongly that bird feeding is a powerful *ethical* issue. This is because I actually care about the birds, about their welfare and conservation. And I am also deeply concerned about the people who are actively involved in feeding birds at their homes, in gardens and on balconies, on a regular basis. Most of these people are committed to the birds that visit, and are motivated by compassion, appreciation and care. They would be devastated if what they were doing proved to be harmful or dangerous for 'their' birds. Many feeders worry constantly about whether the food they provide is okay, but they have no way of finding out. Almost all of them would love to know how to feed their birds better, which food to use and what they should be doing to minimise the risk of genuine issues such as dependency and disease.

My fascination and subsequent obsession with trying to understand wild bird feeding has led me to many unexpected discoveries and terrible truths. Finding out how easily diseases can be spread, that even moderate amounts of some things (mince, sunflowers, cake) can lead to long-term health problems, that the structure of entire bird communities can be changed and how even small amounts of supplementary food can have significant influences on many aspects of birds' lives has been sobering.

On the other hand, my exposure to hundreds of people who feed their birds daily has convinced me of their passion and compassion. Their dedication and concern, commitment and care, have been humbling and uplifting. But this vast population of people who share a common purpose are largely disconnected and unaware of one another. It's like a secret society but without the clandestine meetings. In other parts of the world, feeding enthusiasts eagerly swap stories at their local bird food shops, exchange tips at bird club meetings and record their observations online as participants in hugely popular programs such as *Project FeederWatch* (North America) and *Garden Bird-Watch* (United Kingdom). I keep going on about this stuff, but the contrast with Australia really is extraordinary!

People feeding wild birds in this country have very few places they can go to for ideas and advice, and they are acutely aware that their practice is not something they can talk about openly. This may be regarded as simply unfortunate and at times a little annoying. But it can also be genuinely frustrating when you really need answers – and distressing when something serious happens. Their questions typically include:

- Where do you go to find out whether something is suitable to put on your feeder?
- What is the best mixture of seeds for rosellas?
- Should the birds be eating so much of that mince/sugar/ black sunflower/whatever?
- Is it normal for a magpie to look like that? Could it be related to the food I have put out?
- Should I report these sick birds? To whom? (And what will they say if I admit that I feed?)

I can tell you quite confidently that these types of questions are being asked (privately and usually very quietly) by someone every day in every street in every town throughout the country. And for the most part, there really is nowhere that they can go for answers at the moment. As a result, lots of people remain frustrated and confused. Even more importantly, it is very likely that birds are being exposed to inadequate or inappropriate diets, and heightened risk of disease or injury, all for want of simple information.

IT IS IMPERATIVE THAT WE TRY TO ENGAGE WITH THE PEOPLE FEEDING BIRDS. AND BY 'WE', I MEAN WILDLIFE CARERS, RANGERS, VETS, COUNCILS, SCIENTISTS AND CONSERVATIONISTS.

For these reasons, it is imperative that we try to engage with the people feeding birds. And by 'we', I mean wildlife carers, rangers, vets, councils, scientists and conservationists. Now that I (and you too, I hope) have become aware of these issues, I am compelled to do something. This modest book is just a first step in the much larger project of providing a comprehensive guide to feeding wild birds in Australia. That is going to take a lot of effort over many years. In the northern hemisphere, even after decades of focused research, new details are regularly being discovered and feeding practices are still being refined. We can learn a lot from this work, but the reality is that bird feeding here in Australia is almost completely different; for a start, most of our feeder birds are much bigger, and only a minority are granivorous. I do believe that feeding wild birds in Australia can actually be ethical, responsible and sustainable, and in this book I want to distil some broad principles for those of us who provide food for wild birds, principles that will guide what we can do as concerned feeders. All I can do is start the process – by creating a comprehensive guide to Australian feeding – and hope that other people, organisations and companies will get on board (and many already have).

ABUNDANT BIRDS IN OUR CAPITAL CITIES

These lists show some of the most common species found in our capital cities. Despite the vast distances between these urban centres, you will notice some very familiar names appearing in each city.

ADELAIDE

Australian magpie

Common blackbird

Crested pigeon

Crimson rosella

Galah

House sparrow

New Holland honeyeater

Noisy miner

Rainbow lorikeet

Red wattlebird

BRISBANE

Australian magpie

Australian white ibis

Common myna

Crested pigeon

Magpie-lark

Noisy miner

Rainbow lorikeet

Scaly-breasted lorikeet

Torresian crow

Welcome swallow

CANBERRA

Australian magpie

Common blackbird

Common starling

Crimson rosella

Galah

House sparrow

Magpie-lark

Noisy friarbird

Pied currawong

Red wattlebird

DARWIN

Bar-shouldered dove

Black kite

Dusky honeyeater

Figbird

Helmeted friarbird

Little friarbird

Magpie-lark

Masked lapwing

Rainbow lorikeet

White-gaped honeyeater

HOBART	MELBOURNE	PERTH	SYDNEY
Black-headed honeyeater	Australian magpie	Australian magpie	Australian magpie
Common blackbird	Common blackbird	Australian raven	Australian raven
Eastern spinebill	Common myna	Australian ringneck	Channel-billed cuckoo
European goldfinch	Common starling	Brown honeyeater	Common myna
European greenfinch	Little raven	Magpie-lark	Magpie-lark
House sparrow	Magpie-lark	Pacific black duck	Noisy miner
Little wattlebird	Pacific black duck	Rainbow lorikeet	Pied currawong
New Holland honeyeater	Rainbow lorikeet	Red wattlebird	Rainbow lorikeet
Silvereye	Red wattlebird	Singing honeyeater	Spotted dove
Superb fairy-wren	Spotted dove	Willie wagtail	Sulphur-crested cockatoo

As those of us involved in wild bird feeding await more detailed studies to be undertaken, we should consider the following questions in terms of our practice as feeders.

IS MY FEEDING ETHICAL?

What I am getting at is the fundamental motivation for being involved in feeding in the first place. My studies of and thoughts about feeding over quite a long time have brought me to the conclusion that we all need to be fully aware of the impacts and implications of this practice. It might seem like a harmless hobby, a private pastime that is no one else's business, but we now know that it is much more than that. Improperly practised, feeding can be genuinely harmful to birds, potentially leading to illness, distress, injury and even death. This is serious stuff, so we need to be upfront about why we are doing it.

As an ethically aware bird feeder you should be well informed and conscious of all the issues involved. You will have thought through your personal motivations, and you should be able to answer honestly when asked why you feed birds – your reasons might include simple pleasure and enjoyment,

AS AN ETHICALLY AWARE BIRD FEEDER YOU SHOULD
BE WELL INFORMED AND CONSCIOUS OF ALL THE
ISSUES INVOLVED. YOU WILL HAVE IDENTIFIED
YOUR PERSONAL MOTIVATIONS, AND YOU SHOULD
BE ABLE TO ANSWER HONESTLY WHEN ASKED WHY
YOU FEED BIRDS.

and the hope that your feeding may actually be beneficial to the birds. This awareness might also mean
facing up to the fact that the birds almost certainly
don't need these additions to their natural diet. If so,
what does that mean for the way you feed? Should
you feed them less, more or not at all? There are no
obvious or universal answers to these questions, but
wherever this exercise leads, it should mean we are
much more responsible in the way we feed.

IS MY FEEDING RESPONSIBLE?

So, what does responsible bird feeding look like? For
a start it would involve taking all the issues discussed
in this book very seriously. It will include providing

only small amounts of food, probably just once a day or even less frequently, to ensure that the birds don't become used to something that is not part of their normal diet. A responsible feeder will encourage birds to continue to forage naturally (or maybe just move to the feeder next door ...). Offering small servings reminds us that these visits are supposed to be brief; it is a snack, not a meal. We are not trying to provide all the food the birds need.

Responsible feeding means being acutely aware of the risk of disease. It involves cleaning the feeder thoroughly and regularly, and keeping a close eye on the birds for any signs of illness.

And a responsible feeder will stay as informed as possible about the appropriateness and quality of the food they are providing. There are some simple concepts that can be kept in mind:

- Food must be clean and uncontaminated.
- Don't offer food if it is spoiled or mouldy.
- Avoid any food that contains sugar or salt, or that is too fatty.
- Usually, the less processed the better.
- For packaged products, select those produced by pet or animal food companies.

IS MY FEEDING SUSTAINABLE?

Sustainability can be a slippery concept sometimes, but when applied to bird feeding it refers to ensuring that the practice has negligible impact or influence over the long term. We need to disturb natural cycles as little as possible, but change is always happening. Everyone who feeds birds knows that the numbers and species of visitors can fluctuate daily and seasonally; this is usually related to changes occurring in the bodies of the birds, and in the local landscape. Preparation for breeding, feeding nestlings, the availability of natural foods, long periods of rain or hot weather – each of these will have a strong influence on feeder visits. That is exactly as it should be.

Feeding birds sustainably means being willing – and happy – to be just one component of the birds' normal, natural daily routine. Birds really don't need us as much as we need them.

RESPONSIBLE BIRD FEEDING MEANS PROVIDING ONLY SMALL AMOUNTS OF FOOD; ENCOURAGING NATURAL FORAGING; BEING ACUTELY AWARE OF THE RISKS OF DISEASE; AND STAYING INFORMED ABOUT THE SUITABILITY AND QUALITY OF THE FOOD PROVIDED.

SO, ENJOY YOUR FEEDING – REALLY!

You probably picked up this book expecting a few simple tips on what to feed particular birds and how to do it safely and sensibly. I'll bet you didn't expect so much heavy discussion about all the things that could go wrong. But an awareness of the potentially important issues involved is, I think, essential for everyone engaged in feeding wild birds in Australia. We really are in an unusual position compared with the rest of the bird-feeding world: huge amounts of private feeding despite widespread negative publicity; virtually no accessible or official sources of reliable advice; and very little in the way of safe and suitable commercial feeding products. I have a strong feeling that all this is about to change.

So, let's start this change today, at your place. Start by thinking seriously about why you like to feed. Think about what you should be offering the birds that visit. Ensure that your 'table' is always clean. And then, just enjoy!

ILLUSTRATIONS

FURTHER INFORMATION

1 FIRST

A lot of the details alluded to in *Feeding the Birds at Your Table* are covered in my book *The Birds at My Table*, NewSouth Publishing, Sydney, 2018.

Danielle Hodgson's research on what was included in 'wild' bird food products was part of her Honours dissertation, 'Quantifying Food Composition and Feed Type Preference of Wild Birds Fed in Australia', at Griffith University, 2017.

Andrew Cannon's study was published as: A Cannon, 'The significance of private gardens for bird conservation', *Bird Conservation International*, vol. 9, no. 4, 1999, pp. 287–97.

2 THINK

Excellent discussions on the spread of disease at feeders and dominance interactions can be found in the following articles:

> JS Adelman, S Moyers, D Farine & D Hawley, 'Feeder use predicts both acquisition and transmission of a contagious pathogen in a North American songbird', *Proceedings of the Royal Society* B 282, 2015.

> ML Francis, KE Plummer, BA Lythgoe, C Macallan, TE Currie & JD Blount, 'Effects of supplementary feeding on interspecific dominance hierarchies in garden birds', *PLoS ONE*, vol. 13, no. 9, 2018.

> TE Wilcoxen, D Horn, B Hogan, C Hubble, S Huber, J Flamm, M Knott, L Lundstrom, F Salik, S Wassenhove & E Wrobel, 'Effects of bird-feeding activities on the health of wild birds', *Conservation Physiology*, vol. 3, no. 1, 2015.

The issues mentioned in 'Mince, metabolic bone disease and gapeworm' are discussed in an online article by a vet:

SM Hoppes, *Nutritional diseases of pet birds*, <www.msdvetmanual.com>.

The problems with feeding bread to waterbirds is discussed in the following online articles:

K Mathieson, 'Don't feed the ducks bread, say conservationists', *Guardian*, 16 March 2015.

S Schweig, 'Here's why feeding geese and ducks is bad for them', *The Dodo*, 4 March 2017 <www.thedodo.com/angel-wing-syndrome-geese-ducks-2343009351.html>.

3 FEEDERS

Further details and excellent advice can be found in the following:

Canadian Wildlife Health Cooperative, *Strategies to Prevent and Control Bird Feeder Associated Diseases and Threats*, 2017. Available from: <www.cwhc-rcsf.ca>.

LM Feliciano, TJ Underwood & DF Aruscavage, 'The effectiveness of bird feeder cleaning methods with and without debris', *Wilson Journal of Ornithology*, vol. 130, no.1, 2018, pp. 313–20.

Garden Bird Health Initiative, *Feeding Garden Birds: Best Practice Guidelines*, 2005. Available from: <www.ufaw.org.uk>.

Garden Wildlife Health, *Recommended hygiene precautions*, <www.gardenwildlifehealth.org>.

4 FOOD

Some excellent information on what is good and bad for feeding wild ducks is found in these two online articles:

R McLendon, '3 reasons why you shouldn't feed bread to ducks', *Earth Matters* blog, 2015 <www.mnn.com/earth-matters/animals/blogs/why-you-shouldnt-feed-ducks-bread>.

Melissa Mayntz, *Is feeding ducks bread bad? Can ducks eat bread?*, 2019 <www.thespruce.com/is-feeding-ducks-bread-bad-386564>.

5 HABITAT

Providing habitat – and not just food – in our backyards is an important topic of direct relevance to everyone engaged in feeding birds. In contrast to bird feeding, however, there is lots of excellent and useful information available in Australia. These are just a few particularly good resources.

The *Birds in Backyards* <www.birdsinbackyards.net>program is run by BirdLife Australia and is an essential place to learn all about the birds in your garden, find a huge amount of detailed advice on turning your place into habitat (go to the Creating Places section) and share ideas and pictures on their brilliant Facebook page (Birds in Backyards (Australia)).

Sustainable Gardening Australia <www.sgaonline.org.au>provides a vast amount of information on all aspects of sensible and sensitive gardening in Australia, including how to enhance habitat for birds (see 'Habitat Gardening' under 'Gardening Tips').

George Adams' *Birdscaping Your Garden*, originally published in 1980 and reprinted umpteen times since then, has long been *the* reference book for wildlife gardening in Australia. The latest iteration is bursting with advice and suggestions developed over decades of experience: *Birdscaping Australian Gardens: Using Native Plants to Attract Birds to Your Garden*, Penguin, Melbourne, 2015.

AB Bishop, *Habitat: A Practical Guide to Creating a Wildlife-friendly Australian Garden*, Murdoch Books, Sydney, 2018, is a comprehensive and ecologically informed compilation, full of innovative ideas and sound advice. This is the benchmark for habitat-oriented gardening in Australia.

6 FEED

'Extreme diet switching: seeds and meat': We wrote this up as an article:

> R Gillanders, M Awasthy & DN Jones, 'Extreme dietary switching: widespread consumption of meat by rainbow lorikeets (*Trichoglossus haematodus*) at garden bird feeders in Australia', *Corella*, vol. 41, 2016, pp. 32–36.

THANKS

This little book had its conception in the many lively conversations – in person, online and even by phone – that followed the numerous talks and interviews I was involved in following the publication of *The Birds at My Table*. While these exchanges were usually respectful, what most people wanted was simple, practical, reliable advice. What could they feed their birds?

The problem was that there really was almost nothing out there for them. The organisations and agencies most likely to have information about feeding wild birds at home had been opposing or denying the practice for decades, and could provide little more than pamphlets about dire consequences.

The thing is, there is plenty of important knowledge around. It's just that given the generally negative attitude to bird feeding in Australia, the people in the know tend to keep a low profile. There are many dedicated wildlife vets, pet shop owners, zoo keepers, animal carers, nutritionists and researchers who have accumulated lots of vital information of great value to anyone interested in feeding wild birds. Over the years I have been privileged to meet many of these folk and I have learned an enormous amount from them. They have seen firsthand what can happen when feeding is not done correctly.

There are plenty of people to thank for assisting in the development of this book. Among the most significant were Phillipa McGuinness, who kicked this idea into reality, Emma Hutchinson, who can spot an awkward passage a mile off, and especially Diana Hill, an editor with extraordinary skills and understanding. The team at NewSouth Publishing are a passionate, dedicated and enthusiastic lot and have been a joy to work with; always professional, attentive and supportive. The publishing world can be a complex and mysterious realm but at every step my agent, Margaret Gee, was there, directing, encouraging and getting things done.

I received a remarkable degree of support from some unexpected quarters, including Brisbane City Council, RSPCA Queensland and especially BirdLife Australia. I am particularly grateful for the ongoing and productive cooperation with Monica Awasthy and Holly Parsons from the *Birds in Backyards* program. I also acknowledge the considerable help gained from students and colleagues who collaborated directly and indirectly on many of the issues covered by this book, including Sian Taylor and Rhiann Gilanders and Grainne Cleary (from Deakin University); thanks for your enthusiasm and valued contribution. I would particularly like to draw attention to two students/collaborators from Griffith University: Danielle Hodgson (for extreme dedication and achievement despite serious adversity) and Cara Parsons (for always reminding me that this is all about the animals). This book is dedicated to them.

Numerous knowledgeable specialists shared their views and ideas with me. In particular, I am extremely indebted to three specialist wildlife veterinarians: Tania Bishop, Adrian Gallagher and Robyn Stenner. Despite the controversial nature of this topic, these exceptionally dedicated people were willing to be associated with this project because they passionately hope it changes how people feed birds. A number of others provided information but did not want to be identified, and some were forthright in their profound opposition. I probably should acknowledge that this book is already seen as 'catastrophic', 'ill-informed' and 'likely to cause unparalleled harm' according to some anonymous correspondents. I guess we shall see.

Finally, as always, sincere thanks go to Manon, Caelyn, Dylan and Liz for their patience and good-natured tolerance of my obsessions.

INDEX

Nectar feeder
New Holland honeyeaters

Tube feeder
Spangled Drongo

Seed feeder
Diamond firetail

BACKYARD FIELD NOTES

BACKYARD FIELD NOTES

BACKYARD FIELD NOTES

BACKYARD FIELD NOTES

BACKYARD FIELD NOTES

BACKYARD FIELD NOTES